OF WHALES AND DINOSAURS

© 2016 Lee Kong Chian Natural History Museum

Published by:
NUS Press
National University of Singapore
AS3-01-02, 3 Arts Link
Singapore 117569
Fax: (65) 6774-0652
E-mail: nusbooks@nus.edu.sg
Website: http://nuspress.nus.edu.sg

ISBN: 978-981-4722-13-1 (Paper)
      978-9971-69-855-3 (Case)

National Library Board, Singapore Cataloguing-in-Publication Data

Tan, Kevin, author.
Of whales and dinosaurs : the story of Singapore's Natural History Museum / Kevin Tan.
– Singapore : NUS Press, [2016]
      pages cm
      ISBN : 978-981-4722-13-1 (paperback)
      ISBN: 978-9971-69-855-3 (case)

      1. Lee Kong Chian Natural History Museum. 2. Natural history museums
– Singapore. I. Title.

QH70
508.0745957 -- dc23   OCN900026940

With the Support of 金基氏李
LEE FOUNDATION

Cover artwork by: Ernest Goh, www.ernestgoh.com
Designed by: Nelani Jinadasa
Printed by: Markono Print Media Pte Ltd

Opening spreads: 0.1 The 'Singapore whale' hanging from the roof of the original home of the Raffles Museum at Stamford Road (photograph ca. 1960s); 0.2 Apollonia, one of the three giant sauropod dinosaurs in the new gallery.

OF WHALES AND DINOSAURS

THE STORY
OF SINGAPORE'S
NATURAL HISTORY
MUSEUM

KEVIN Y.L. TAN

NUS PRESS
SINGAPORE

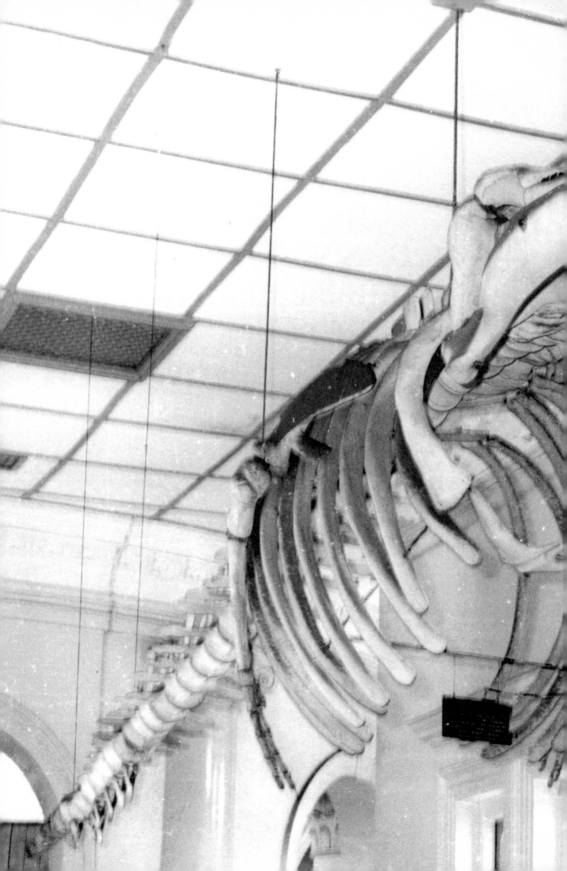

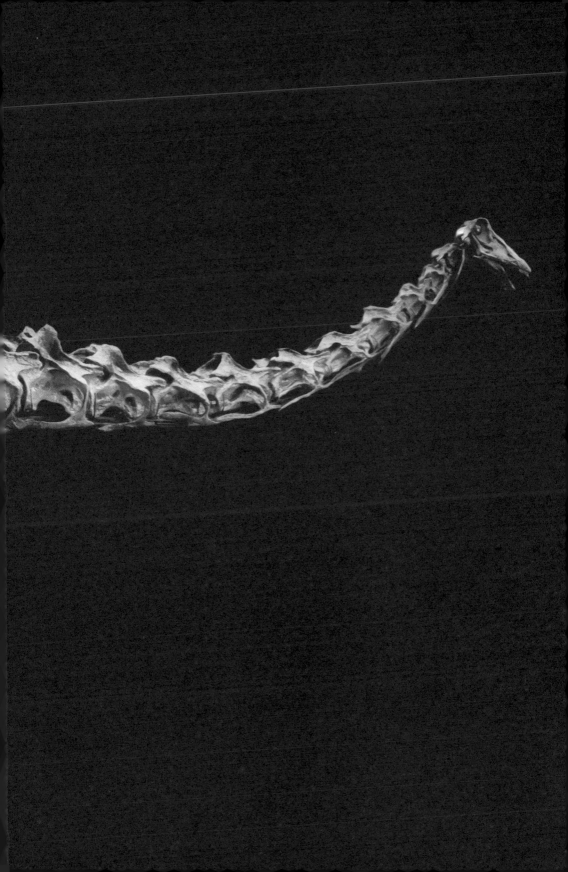

Contents

ACKNOWLEDGEMENTS

THE WRITING OF this book was, for most of us, a labour of love. Many individuals gave so generously of their time and energies in helping me bring the incredible stories of the Collection and of the Museum—in its various incarnations and guises—to life. My first thanks goes to Professors Peter K. L. Ng and Leo W. H. Tan who approached me to write this book in full trust and belief that I would be able to tell the story 'honestly, warts and all'. One could not ask for better 'bosses' for in the true spirit of academic freedom, they left me entirely to my devices on how best to get the job done. It was a great pleasure working with them.

Professor Peter Ng's staff at the Raffles Museum of Biodiversity Research was also extremely helpful in providing material, photographs, leads, and other materials not found in the public domain, and for arranging for a number of interviews with key individuals. Thanks to Dr Tan Swee Hee, Mr Kelvin Lim, Dr Tan Heok Hui, Mr Martin Low, Ms Belinda Teo, Ms Greasi Simon, Ms Shu Shwu Li, Mr Marcus Chua, Dr Joelle Lai and Ms Toh Chay Hoon.

I would also like to thank the following individuals who assisted me in my research, and in helping me acquire 'hard to get' materials: Ms Fiona Tan shared with me her research material which she gathered for the Museum and which she also used in her BA Honours thesis, *Natural History in the Raffles Museum*; Dr Tan Swee Hee and Mr Martyn Low handed me five boxes of materials belonging to the museum which I could use; Ms Tan Teng Teng, spent much time scouring the newspaper archives for relevant materials; Ms Angela Gosling, spent hours photographing records at the archives of the Natural History Museum in London; and Ms Louise Macul, who sent me interesting material from the Sarawak Museum. Thanks too to Mr Eric Chin, Director of the National Archives of Singapore for facilitating my access to the Ministry of Culture files, and to his helpful staff, especially Ms Yao Qianying.

The story of the museum is that much richer, thanks to the face-to-face interviews I was able to arrange. My eternal thanks to the following interviewees (in alphabetical order): Mr Eric Alfred, Dr PN Avadhani; Ms Nancy Byramji @ Aurora Hammonds; Professor Lam Toong Jin; Professor Lee Soo Ying; Ms Anna Sharma; Professor Bernard Tan Tiong Gie; Professor Leo Tan Wee Hin; Dr David R Wells; and Mrs Yang Chang Man. I am particularly grateful to Ms Ilsa Sharp who spent much time linking me up with various individuals who were connected with the saving of the Collection, including Nancy Byramji and PN Avadhani.

Finally, my thanks goes to the two anonymous reviewers who offered numerous excellent suggestions on how to improve the manuscript, as well as to Mr Peter Schoppert and his team at NUS Press in producing the volume. Special thanks goes to Ms Christine Chong, who helped pull the whole book together, and to Clarence Ng, whose excellent fact-checking prevented me from committing some egregious factual errors. That said, all errors and foibles remain mine alone.

Kevin Y. L. Tan
February 2015

Why a Natural History Museum?

Originally *musæum* had two definitions. It was most traditionally the place consecrated to the Muses (*locus musts sacer*), a mythological setting inhabited by the nine goddesses of poetry, music, and the liberal arts ... More specifically, *musæum* referred to the famous library at Alexandria which served as a research centre and congregating point for the scholars of the classical world.

—"The Museum: Its Classical Etymology
And Renaissance Genealogy," Paula Findlen

In 1972, Singapore nearly lost an invaluable piece of the past when its natural history collection was under threat. The collection had for many years formed the core of the only omnibus museum in Singapore: the Raffles Museum, later the National Museum. It is particularly fitting that in its Golden Anniversary—when there are museums for just about everything from art to sports and even toys—Singapore will see this collection housed in the new Lee Kong Chian Natural History Museum, an independent institution dedicated to natural history.

Like all museums, a natural history museum has three primary missions: to build collections, facilitate research, and educate. However, it differs from all other museums in one major respect: its subjects are not the products of human endeavours but of evolution. They are the products of a billion years of unrelenting and merciless natural selection at work. The natural history museums of old were happy to display taxidermised and mounted specimens of all kinds—a zoo of the dead. Collectors of nature assembled strange plants and animals to excite and arouse the curiosity of scholars and the public alike. At that point, visitors were content just to be awed by these dead but wonderful specimens.

That was Raffles' vision when he mooted the idea of a museum in Singapore. Formally established as the Raffles Library and Museum in 1878, Singapore's natural history museum began life at Stamford Road in 1887 exhibiting preserved animal specimens from Southeast Asia. Although the original Raffles Museum had a strong zoological slant, it also featured exhibits of early human culture, anthropology and geology.

At around the same time the old Raffles Museum was set up, its two famous "brothers" were also established in the region: the Sarawak Museum in 1888 (also started by the British) and the Museum Zoologicum Bogoriense (or Bogor Museum) in Java in 1894 (by the Dutch). The Sarawak Museum's original building in Kuching is still intact and being used; but there are ambitious plans to grow it. The Bogor Museum is a huge instititution with millions of specimens and hundreds of researchers, beautifully nestled in the grounds of the famous Bogor Botanic Gardens; and with modern storage and research facilities in Cibinong, just outside Bogor. This trio are the earliest museums of their kind in Southeast Asia.

The natural history museum as an institution became more sophisticated as the scientific field advanced. It took men like Charles Darwin and Alfred Russel Wallace to ask deeper, more probing questions. (A historically valuable flycatcher contributed by Wallace is now held in the museum, see p. 28.) The most powerful was the simplest: why are there so many species? Out of this question came the field of evolutionary biology, a discipline that asks profound questions: how life has changed, why it has changed, and how it affects things around it.

Today, natural history museums hold specimens that are a splendid testimony to the myriad forms of life on Earth. The preserved tissues, dehydrated DNA and shrivelled biostructures are the result of millions of years of evolution; of success reinforced by the cauldron of natural selection. They are a library of information past and present: what works, what does not, a beacon to our own future. We have much to learn from the dead. The millions of specimens in the world's museums are raw material for evolutionary scientists and naturalists.

A key part of this legacy and repository—more than 500,000 specimens—is in Singapore's natural history museum. Some specimens housed here are so valuable that they must be guarded jealously.

In a few very rare instances, uncommon specimens may need to be sacrificed. The museum had a unique three-eyed crab—an anomaly of nature and the only one on the planet. However, to understand how it became so odd, a renowned German developmental evolutionary biologist had to section it. The crab is now in the museum in a box of a dozen slides. Unless specimens are used, they have no place in a modern natural history museum.

While historians may be interested in the Wallace specimen and the layman fascinated by anomalies, scientists and researchers are excited by what seems more prosaic: type specimens. A holotype is *the* specimen used by the original scientist to name and describe a particular species and serves as the ultimate representative of the species. When anyone had doubts, queries or debates on the validity of a species, what its name should be etc., they make a decision based on the holotype. Equally important are paratypes, or specimens used at the same time as the holotype to name a species.

Natural historians hunt down strange and new species, new ecosystems, and unexpected adaptations. Like the Starship Enterprise, they 'explore strange new worlds [and] seek out new life and new civilizations,' and 'boldly go where no man has gone before.' Staff of the museum have scaled lofty mountains, traversed headwaters of some of the world's greatest rivers, and plunged into limestone caves cloaked in darkness. Sometimes, they take weeks just to find a particular rare species. Modern naturalists must be more than good scientists; they must be excellent hunters. Tough and adaptable, they must be able to climb, swim and dive. It is hard work just to find and get the specimens, then collate the associated data (provenance, habitat, ecology, photographs, DNA etc.) needed for subsequent research. Today, naturalists need to handle and study animals or plants 'in the flesh'. The Lee Kong Chian Natural History Museum considers as its prized assets its over 11,000 type specimens representing over 1,240 species and subspecies from throughout the region.

But the fates of natural history museums are not always in the hands of scientists. As Singapore entered the 1970s, the government did not see any reason to keep the zoological collection as economic development and industrialisation took centre-stage. Had it not been for the unstinting efforts of a few stout-hearted individuals, this collection—the largest collection of Southeast Asian animals—might well have been lost, its existence forgotten. Thanks to the vision, fortitude and efforts of past directors of the old Raffles Museum, three of whom I had the privilege to know and work with—Michael Tweedie, Eric Alfred and Yang Chang Man—Singapore has managed to retain its natural history treasures. The museum morphed into the

Zoological Reference Collection in the 1970s and the Raffles Museum of Biodiversity Research in 1998, all the while still focused on regional natural history and improving its collections.

The Lee Kong Chian Natural History Museum today is truly built on the foundations laid by others and is committed to continue studying the biodiversity of Southeast Asia. But all their sweat will be in vain if there is no future for the next generation of scientist-naturalists in Singapore. Biodiversity research is ultimately a global challenge. Communities and individuals in the field must include Singaporeans who will help safeguard the collections, advance the research and take education to new heights; those from the region who will use their experiences to help their own countries; and international researchers who will use their connections to bring the scientist-educator-manager closer together.

There are other important natural museums in the region: the National Museum of the Philippines in Manila (1901); the Sabah Museum in Kota Kinabalu (1965); and the National Science Museum outside Bangkok (1995). There are also a large number of state, university or research institute-based and even private museums with key specimens. The Lee Kong Chian Natural History Museum stands out in not only being a museum of Singapore's biodiversity, but also as a one-stop facility for the study of the region's richness. This is a matter of great pride and importance because at the end of the day, biodiversity science is not and cannot be restricted and confined by geopolitical boundaries.

The potential for education and enthusing the public about natural history is also enormous. Firstly, since most people already have a general interest in nature—and can appreciate a beautiful flower, gasp at the sheer size of a dinosaur, and shudder at a fearsome crocodile— natural history museums cut through culture, language and nationality to attract visitors. The second level of interest is practical: thousands of students at local tertiary institutions are exposed to the natural sciences each year and the museum's specimens are an invaluable resource for young researchers and professional scientists interested in the biodiversity of Southeast Asia.

Finally, the collection fuels the sheer passion that many Singaporean 'citizen scientists' have for natural history. Such amateurs draw on natural history and reference the museum's material for their studies, pursuing this not for fame or fortune but for pure enjoyment and the desire to make a difference in local and regional conservation. Some individuals may even become world-renowned authorities in their interest areas! The continuing expeditions, surveys and excellent network of collaborators in the region mean that young scientists and natural historians have a wonderful chance to experience many new ecosystems, and all manner of strange plants and animals.

That this wonderful *grande dame* of museums is now reestablished as the Lee Kong Chian Natural History Museum, with its original mandate for research and education, is a fantastic thing in itself. But a natural history museum is also a sign that the nation has come of age in its fiftieth anniversary: we have managed to carve up a space in pragmatic Singapore, and in our lives, for something as ethereal as nature and heritage. The myriad variety of life forms are found, preserved and catalogued not because they have commercial value. In fact, the specimens can be considered worthless in terms of monetary value. But from a scientific and educational perspective, these specimens are priceless.

At the same time, viewing these exhibits—many specimens that are long dead and no longer with us—is a form of time travel. Plants and animals that lived in our country's youth were eradicated without a second thought. In fact, up to a third of the species that were found in Singapore are now extinct. Their preserved bodies remind us of their previous presence, our ecological folly, and mistakes we should never repeat. In short, a museum is a reflection of our past, a vessel for our memories, and a reminder of what we are. If 'what's past is prologue', then what is to come will surely be exciting for all of us.

From the Raffles Museum to the Lee Kong Chian Natural History Museum

1874–2014

THIS IS THE story of the Zoological Collection that lies at the core of the Lee Kong Chian Natural History Museum. Over the years, this very special and historic collection has been known by different names. It originated from the Singapore Museum—renamed 'Raffles Museum' in 1874 and 'National Museum' in 1960—an omnibus museum with collections of natural history, anthropology and art. In 1972, after it split from the National Museum, it was often referred to as the 'Raffles Collection' or the 'Raffles Natural History Collection.' After it became safely ensconced in the Department of Zoology at the National University of Singapore, it became known as the Zoological Reference Collection, and from 1998, formed the core of the Raffles Museum of Biodiversity Research at the National University of Singapore.

The history of this collection goes back to the formation of the Singapore Museum in 1849. This story inevitably overlaps with the histories of many other institutions of Singapore life. These other stories provide the necessary background to our story but are only tangentially important to the story of how this collection came to be, grew, struggled, and ultimately found a permanent home in the new Lee Kong Chian Natural History Museum.

Because of the way the collection has been displayed and discussed, people are confused about what it refers to exactly. Part of the reason stems from the fact that the Zoological Collection is made of two types of specimens and is actually in two parts—the mounted specimens (the display collection) and the scientific specimens (the reference collection). Most people who have visited the old Raffles Museum or National Museum prior to 1972 will remember the many animal exhibits—the overarching mounted whale skeleton, tigers, crocodiles, insects, etc.—that were scattered throughout the various galleries of the museum. While these exhibits were the most famous and visible signs of the museum's zoological collection, they were only part of the display collection. These specimens were mounted and prepared for exhibition and sought to give visitors an idea of what such an animal or bird might look like in real life. They were not meant for scientific study. Over the years, the museum obtained, taxidermised, exhibited and discarded thousands of such mounted exhibits. They were visually impressive and provided the visitor with the most palpable experience, but were generally of little scientific interest to the scientists.

Scientists and zoologists are far more interested in the specimens in the reference collection because of the raw data they could potentially extract from their examination. Better yet if these specimens were holotypes (the very first specimens used to describe new species) or paratypes (specimens of types other than the holotype). Reference collections of any museum or scientific institution are hardly ever put on display as reference specimens were preserved as skins, skulls or kept in bottles with formalin or alcohol. Indeed, curators fear that specimens that are too heavily exposed, handled, or displayed may deteriorate beyond repair and lose their scientific value. It was no different with the Raffles Zoological Reference Collection. Most of the time, they lay quietly hidden in boxes, cabinets and drawers in the storerooms deep in the recesses of the museum or at the bottom of display cabinets. When the reference collection was cleaved from the National Museum's general inventory in 1972, no one had any idea exactly how large this collection was. The fact that it took the staff almost a year to list and move the collection should give some idea on the huge amount of scientific material that had been accumulating at the museum since 1849.

The relative invisibility of the reference collection and its esoteric appeal to scientists and higher-level students of science meant that the general public had no particular affinity for it. And as will be seen in the accounts that follow, and especially of efforts to save the collection, journalists happily juxtapose the specimens in the reference collection with the display collection only because the latter are so much more instantly recognisable and hence lovable. Few members of the public have seen the thousands of bird skins in the reference collection. And even if they had, they are not likely to feel the same way they do about them as they do about the display specimens. The much lamented whale skeleton, which the Science Centre gave to the Muzium Negara in 1974, still haunts the minds of those who saw it. This lack of visibility does not, however, make the story any less compelling. The Zoological Reference Collection is the closest thing Singapore has to a national treasure.

This book starts off with an account of the founding of the Singapore Library and Museum, and how it grew and was transformed into the Raffles Museum and Library. These early years—from 1849 to the 1920s—were crucial for the formation of the institution and the creation of a museum-going culture amongst the local inhabitants in Singapore. In these years, the display specimens took pride of place in the museum; everyday, people would trample through the museum to be awed by them. The reference collection, however, only really got started in the 1920s with the curatorial careers of three remarkable zoologists—Herbert Christopher Robinson, Cecil Boden Kloss and Frederick Nutter Chasen—who were the giants of the field in the inter-war years (1919–42). It was Robinson and Kloss who arranged for the enormous mammal and bird reference collections of the Selangor Museum to be exchanged with the entomological collection of the Raffles Museum in 1927; this established the Raffles Museum as the leading reference collection in the region.

However, it was the first two decades of Singapore's independence that proved to be the toughest years for the museum and for the collection. It was then abandoned by the Singapore Government in 1972 because it was considered irrelevant and had little to contribute the all-important nation-building processes of the time. For years, it lingered, forgotten by Singaporeans. These were precarious decades for the collection and the fact that they survived, and continued to be cared for in the midst of the most appalling conditions, highlights the heroic efforts of the few individuals who saved Singapore's natural history heritage from oblivion. Indeed, had it not been for the dedication and tenacity of a few stalwarts who really understood the value of the collection, it would have been lost forever. A large part of this book is dedicated to those who worked tirelessly to keep the collection in Singapore and intact.

CHAPTER ONE

Temenggong Ibrahim's Coins and the Founding of the Raffles Museum

IN 1987, SINGAPORE'S National Museum celebrated its centenary with an art exhibition and a whole host of other events. A limited edition $5 coin was struck, and a three-stamp first-day cover was issued to commemorate the centenary. The natural conclusion would have been that the museum must have been founded in 1887. It was not.

Its history goes back almost forty years earlier when the Singapore governor presented two ancient Achinese coins—given by Temenggong Daing Ibrahim—to the Singapore Library. And if we are to consider its very genesis, the idea of establishing a museum in Singapore goes back to 1823 when Sir Stamford Thomas Raffles himself founded the Singapore Institution. Legally, the museum was established by the passage of the Raffles Societies Ordinance in 1878.[1] Thus, the anniversary celebrated in 1987 was merely the centenary of the National Museum building on Stamford Road, not of the museum itself.

A Rafflesian Dream

The story of the National Museum and its collection begins with the inception of the Singapore Institution in 1823. Sir Thomas Stamford Bingley Raffles (1781–1826) was a visionary and a dreamer. One of his abiding dreams for Singapore was that an educational institution be established in which young men of Singapore and Malaya would be educated not only in their own languages and cultures but also in the best English tradition. In *Minute on the Establishment of a Malay College in Singapore*,[2] he envisioned a full-fledged college, with departments of Chinese, Malay and Siamese, alongside that of English, to prepare young men to be future leaders of Singapore. And among other things, this college would have a splendid library and museum. Given that Singapore was such a small settlement where even basic formal education was non-existent, it was an ambitious but unrealistic plan.

During his last visit to Singapore in 1823, Raffles convened a meeting to establish an educational institution along the lines of his vision. Among those present at the meeting, held on 1 April 1823, were Raffles; his good friend, Dr Robert Morrison; Reverend Robert Sparke Hutchings (the chaplain of Penang and founder of Penang Free School); Sultan Hussein; and Temenggong Abdul Rahman. Raffles donated $2,000 (Spanish dollars) and subscribed $4,000 on behalf of the East India Company. His wife Sophia gave $400 while Dr Morrison contributed $1,200. The Resident, Lieutenant Colonel William Farquhar, Sultan Hussein and Temenggong Abdul Rahman each contributed $1,000. In all, $17,495 was collected at that first meeting. A Board of Trustees headed by Raffles was constituted.

The trustees resolved to erect a building on a site at Bras Basah Road as the premises for the new Singapore Institution. Lieutenant Philip Jackson (1802–79), who was then Executive Engineer and Surveyor of Public Land, told the Board of Trustees that he could complete the building within 12 months for $15,000 (Spanish dollars). On 5 June 1823, the foundation stone for the Singapore Institution was laid. Despite the auspicious start for the Institution, the road ahead was fraught with troubles. By December 1823, the Trustees and Lieutenant Jackson had run out of money to complete the building. Funds from the proposed sale of Dr Morrison's Anglo-Malay College in Malacca did not materialise. The building remained unfinished for the next 15 years. In the meantime, Raffles had died in 1826, his dreams of a great institution of learning in Singapore unfulfilled.

A Fresh Start

In 1836, a new Committee of Trustees engaged Irish architect George Drumgold Coleman—who was also Superintendent of Public Works—to complete the building and add a new wing to it for $5,700. Both were completed by December 1838. The new wing, which had been added towards the Stamford Road end of the compound, had three classrooms on the ground floor and three on the upper level. This would later serve as the boarding house of the school. When these works were completed, the committee commissioned a third wing to be built (this is the one closest to Bras Basah Road). At a meeting of the trustees of the Singapore Institution in May 1836, it was decided that they would establish an elementary school instead of a tertiary institution, which was Raffles's original plan. This was a much more realistic and practical plan since it was what the fledgling settlement most urgently needed.

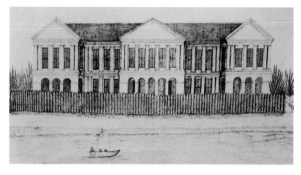

1.1 The earliest known drawing (1841) of the Singapore Institution, a pencil-and-ink and watercolour sketch by J. A. Marsh

The buildings were only occupied in December 1837, when students of the Singapore Institution Free School moved in. They were later joined by the Singapore Free School. Reverend Fred J. Darrah, Chaplain of the Mission Chapel in Singapore, had established Singapore's first functioning school when he founded the Singapore Free School on 1 August 1834. It was organised under the auspices of the Singapore School Society and initially had 46 students. This number swelled to 75 within two months. Before moving into the Stamford Road premises, it occupied an attap-roofed building in High Street, near the junction with River Valley Road. In addition to transforming the Singapore Institution into an elementary school, the Trustees also proposed

> …to appropriate one of the upper rooms, as a Library and Museum, and where all meetings of the Committee, or of the Trustees may be held. To form the Library and Museum, donations of books should be forthwith solicited—as also specimens of the Natural History of the Archipelago and the countries in our vicinity. If only a little zeal be displayed in accomplishing these two desirable objects, collections would soon be made, which would form perhaps some of the principal attractions of the Institution after its completion.[11]

It is interesting that although the Committee of Trustees had jettisoned Raffles's original plan, it maintained the desire for a library and museum. Plans were drawn up for such a library in 1841 but we have no record as to whether it was actually furnished and stocked with books. Indeed, it was not till January 1845 that a private library, called the Singapore Library, was officially established. William Napier (1804–79), Singapore's first law agent, chaired the meeting and made 'a handsome donation of more than a hundred volumes, consisting of

1.2 Portrait by George Francis Joseph (1817) of Raffles after he was knighted by the Prince Regent

Raffles the Naturalist[3]

Other than founding of Singapore in 1819, Raffles also founded the Zoological Society of London,[4] a few months before his death in 1826. Natural history had long been one of Raffles' abiding interests. Recounting Raffles' early life, Lady Sophia Raffles wrote: 'As a school-boy, his garden was his delight; to this was added a love of animals, which was perhaps unequalled.'[5]

As an adult, Raffles was only able to indulge in his passion for natural history after he left London for Penang. It was recorded that while he was in Penang between 1805 and 1810, he kept 'a number of animals in cages, including a siamang.'[6] In 1811, while in Malacca preparing for the invasion of Java, Raffles reportedly employed four men to search for natural history specimens for him. Abullah bin Abdul Kadir (more famously known as Munshi Abdullah) wrote this about Raffles' deep interest in natural history:

> He kept four persons on wages, each in his peculiar department: one to go to the forests in search of various kinds of leaves, flowers, fungi, pulp, and such like products. Another he sent to collect all kinds of flies, grasshoppers, centipedes, bees, scorpions, giving him pins in a box to put through the creatures. Another he sent with a basket to seek for coral, shells, oysters, mussels, cockles, and such like: also fishes of various species: and yet another to collect animals, such as birds, jungle fowl, deer, stags, mousedeer and so forth. Then he had a large book with thick paper, whose use was for the keeping of the leaves and the flowers. And, when he could not put them there, he had a Chinese Macao painter, who was good at painting fruit and flowers to the life, these he sent him to copy. Again he kept a barrel of arrack or brandy, and when he got snakes, scorpions, centipedes and other such like, he would put them in till they were dead, before putting them in bottles. This occupation astonished the people of Malacca, and many profited from going in search of the living creatures that exist in the sky and the earth, sea or land, town or country.[7]

During the time he was in Malacca, Raffles engaged in a contest with Major William Farquhar to discover new animal species.[8] Farquhar, who had been in the Straits for far longer than Raffles had, discovered the Malayan Tapir (*Tapirus indicus*) and the bearcat or binturong (*Arctictis binturong*). Raffles claimed to have already known of these mammals earlier and tried to have his description published and Farquhar's withdrawn. In the case of the tapir, a five-year delay in publication meant Farquhar lost out to a young French zoologist Fierre Diard, who published his description a year earlier; however, he generously stated that Farquhar had discovered the species.

Later, when Raffles was Lieutenant-Governor of Bencoolen, he had a Malayan Sun Bear (*Helarctos malayanus*) which 'was brought up in the nursery with his children and was allowed to sit at his table where it showed a marked taste for champagne.'[9] Indeed it was Raffles who first described this species of bear in 1821.

Raffles' deep interest in natural history was heightened by his interactions with some of the leading natural scientists then studying the region. One of the first was the American naturalist Thomas Horsfield whom Raffles first met at Solo (modern day Surakata) towards the end of 1811. At the time, Horsfield already spent a decade researching the natural history of Java under the Dutch colonial administration. Raffles was anxious to secure from Horsfield his collection of insects, mammalia and birds for the East India Company in London. Raffles succeeded and deposited the collection in the Company's museum in London.

CALYPTOMENA VIRIDIS.

1.3 Plate from Thomas Horsfield's Zoological Researches in Java
(1822). The Green Broadbill *(Calyptomena viridis), now extinct on the
island, was described by Raffles when he was in Singapore in 1822.*

According to John Bastin, Raffles' interest in natural history intensified during his 1816–17 visit to London where he met with London's leading naturalists, including Sir Joshua Banks and Dr Joseph Arnold. Together with the latter, Raffles later discovered the world's biggest flower, the carnivorous Rafflesia (*Rafflesia arnoldi*) in 1818. Alas, life was very hard for naturalists in the tropics and Arnold died only four months after arriving in Sumatra. After failing to persuade the Company to send a replacement, Raffles met Dr William Jack through Dr Nathaniel Wallich, the Danish superintendent of the Botanic Garden in Calcutta who was also surgeon and botanist. Jack was a talented naturalist and although he too, died prematurely, aged 27, in 1822.

Raffles also employed two French zoologists, Pierre-Médard Diard and Alfred Duvaucel, who amassed a large zoological collections under Raffles' patronage. According to Bastin, the two Frenchmen failed to honour an agreement to publish their discoveries in England, and this 'forced Raffles to seize their collections and undertake a scientific description of them himself.'[10] Raffles prepared a *Descriptive Catalogue* to accompany the collection which he shipped to London in 1820. Although the collections were essentially those assembled by Diard and Duvaucel, and the catalogue itself largely the work of William Jack, it established Raffles's reputation as a zoologist.

During Raffles' last visit to Singapore in 1822–3, he met up with Wallich, who was on the island on sick leave. Raffles joined him in botanical excursions on the island and was encouraged by Wallich to establish a botanical garden in Singapore. This he did by allocating a piece of land on Government Hill (later Fort Canning) for this purpose. Raffles returned to Bencoolen in July 1823 with 'a variety of natural history collections from Singapore' and others from Sumatra. Alas, this material was all to perish in the tragic fire that engulfed his vessel, *Fame*, in February 1824.

Bastin tells us that Raffles' interest in natural history straddled its whole range from geology, to ichthyology and entomology. The curtailment of his political activity following the establishment of the settlement in Singapore in 1819 were perhaps, Bastin says, 'the happiest period of Raffles' life when, after assiduous study and close associations with a number of naturalists over many years, he enjoyed the necessary leisure to engage in natural history research for himself.'

1.4 *The* Rafflesia arnoldii, *the world's biggest flower, which is named in honour of Raffles and his friend Joseph Arnold (1782–1818) even though it was first discovered by French explorer Louis Auguste Deschamps (1765–1842) in Java in 1797. Raffles and Arnold collected a specimen of this plant in Sumatra in 1818.*

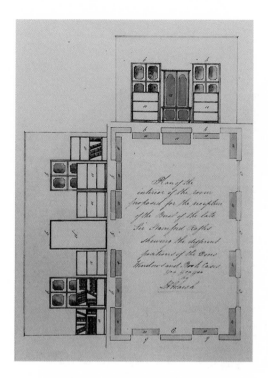

1.5 Plan of the Library of the Singapore Institution by J. A. Marsh, dated 1841. It is not known if this plan was executed.

nearly complete sets of the *Edinburgh and Quarterly Reviews*.[12] The library was to be located in the 'large and commodious' Library Room that was provided gratis by the Singapore Institution.[13] Part of Raffles' vision was beginning to materialise. Perhaps, as a sign of the realisation of Raffles' dream, his marble bust by the well-known London sculptor Sir Francis Legatt Chantrey—which Lady Raffles presented the Singapore Institution in the 1830s but had somehow 'found its way to the Court House'[14]—was moved to the library in 1846. The library did well, and by 1846 was described to be 'in such a flourishing condition, and there seems now no reason to doubt of its continued prosperity and permanence'.[15]

Over the next three years, the library appeared to function satisfactorily, attracting more subscribers and donations of books. However, plans for also incorporating a museum on its premises seemed to have been temporarily shelved.

The Birth of the Museum: The Temenggong's Gift

On 26 January 1849, Governor William John Butterworth wrote to H. C. Caldwell, Chairman of the Singapore Library, requesting that the gift of the two coins from Temenggong Ibrahim be placed in the Reading Room:

Gentlemen,

It has been suggested by JR Logan Esq,[16] to whose exertions we are indebted for the present most promising Library, that the accompanying coins would be appropriately placed in the Reading Room of that Institution, as the nucleus of a Museum, tending to the elucidation of Malayan history, which it is hoped may eventually be formed at this station.

Under these circumstances, and with the above view, I beg to present the coins in the name of His Highness, Sree Maharajah, the Tamoongong of Johore, who purchased them from the Convicts employed in constructing the Road to 'Teluk Blangah' or New Harbour, in the vicinity of which they were discovered about 8 or 9 years since.

JR Logan Esq, to whom these coins have been submitted, observe, that they are 'not the coins of Johore but probably Achinese of whose invitation I found many traces and traditions, up the Johore River' that the Inscriptions are as here noted…

It will afford me much pleasure to acquaint His Highness, the Tamoongong with your acceptance of the coins, for deposit in the Singapore Library Room.[17]

The Managing Committee of the Library met three days later and unanimously decided to accept the Temenggong's gift. Caldwell acknowledged the Governor's letter and informed him that at a meeting of the Committee convened that very same day, 'it was resolved unanimously, that the above mentioned coins should be accepted and deposited in the Library Rooms' and that the Committee intended to propose the establishment of 'a Museum in connection with this Library' at the annual meeting of the proprietors on 31 January 1849.[18] Caldwell, seconded by L. Fraser, proposed that

…a Museum with a view principally to the collection of objects to illustrate the General History and Archaeology of Singapore and the Eastern Archipelago, be established in connection with the Singapore Library; that it be called the 'Singapore Library-Museum' and that it be deposited in the rooms of the Library.[19]

This proposal was unanimously accepted by the proprietors of the library and a committee comprising Caldwell, James Richardson Logan, Abraham Logan, Thomas Oxley, H. Man and W. Treil was formed to framed rules 'to regulate the Museum and to procure contributions of object for the Museum.'[20] Two months later, a set of 'General Rules for the Museum' was passed. Rule 1 stated that the museum would be called the Singapore Museum and that it would principally acquire a 'collection of objects to illustrate the General History and Archaeology of Singapore and the Eastern Archipelago.' Having now established the museum, the proprietors were anxious to ensure that it would have to be a permanent fixture in the settlement; they declared in Rule 7 that the museum was to be 'indissoluable, and shall not be removed from the Settlement.'[21]

1.6 The Singapore Town Hall, where the Raffles Museum was housed from 1874 to 1876

Contributions and donations from the public would be sought for the following items:

1. Coins; 2. Manuscripts; 3. Inscriptions on Stone or Metal; 4. Implements, cloth or other articles of native art or manufacture; 5. Figures of deities used in worship; 6. Instruments of war or other weapons; 7. Instruments of Music; 8. Vessels employed in Religious ceremonies; 9. Ores of metals; 10. Minerals of every description; 11. Fossils; and any other object which may be considered suitable for the purposes of a Museum.[22]

To this end, an article appeared in the *Singapore Free Press* on 7 March 1850 soliciting donations and explaining the objects of the museum. It was noted that 'the Museum is now a most valuable addition to the Library, and doubtless so esteemed by its visitors, who are thus further attracted to it by a medium so interesting.'[23] The following year, the official Museum Committee comprising Thomas Oxley (Chairman), James Richardson Logan, Abraham Logan, H. C. Caldwell, G. W. Earl, Robert Little, J. Harvey (Treasurer) and J. C. Smith (Secretary), was elected into office. As the future Museum Director Richard Hanitsch observed, 'there are only few clues to what sort of collections the Museum contained at the time.'[24] We do know that some Malay tombstones from Malacca were donated to the museum in 1852, and that 'Native Arms, illustrations of Natural History, and other suitable articles' were presented to the museum in 1854, as was a portrait of Sir James Brooke in 1855.[25] Unfortunately, this is the last we hear of the museum for the next 18 years. Annual Reports of the Singapore Library became increasingly sporadic and there was no mention of the museum till January 1874.[26]

In the meantime, the Singapore Library, which had moved to the Town Hall (now Victoria Theatre) in 1862, was 'discovered to be in financial difficulties.'[27] The lack of new subscribers meant that funds for the purchase of new books and periodicals dwindled as the years went by. At the same time, the library appeared to have suffered from poor management, became mired in debts, and by the end of 1873, was in danger of shutting down. We can only surmise that beyond the passing of 'Rules for the Museum' and the safekeeping of Temenggong Ibrahim's coins, nothing in the way of a serious museum materialised in those years.

Metropolitan Influences

On 13 May 1873, a despatch from the Secretary of State for the Colonies, the Earl of Kimberley, was laid before the Legislative Assembly. This despatch contained a proposal 'to establish and maintain, at a reasonable cost to each Colony, a permanent Exhibition of Colonial Products in connection with the Exhibition Building of South Kensington.'[28] Items requested included not only commercial products but also 'objects of interest of whatever kind, illustrating the ethnology, antiquities, natural history and physical character of the country.'[29] This matter was deliberated at a meeting of the Legislative Council on 4 June 1873 at which Thomas Scott of the Singapore Chamber of Commerce pledged the Chamber's support for the proposal. On the motion of the Colonial Secretary, the Council resolved to contribute a sum of £404 9s 6d (pounds, shillings and pence) for the construction, and an annual subscription of £21 3s 1d for the maintenance of the proposed Exhibition, but nothing appears to have been done to organise and curate the exhibition.[30]

In December that year, Dr H. L. Randal, Principal Civil Medical Officer of the Straits Settlements wrote to the Colonial Secretary proposing that necessary steps be taken 'to establish in Singapore a Museum for the collecting of objects of Natural History.'[31] Randal opined that such a collection

...might in time be easily procured which would not only be of immense value and interest to the scientific world, but would also afford a general interest to the residents of the Straits, and be a work in which the government might fairly anticipate every encouragement and assistant from our own Community and those around us.[32]

Governor Sir Andrew Clarke, pleased at Randal's initiative, replied with the suggestion to combine the museum with a public library.[33]

1.7 *Governor Sir Andrew Clarke, who initiated the Museum of Natural History in Victoria, Australia, some 20 years earlier, was keen to combine a museum with the library.*

A Public Library and Museum

By the early 1870s, Singapore was a thriving, prosperous settlement with a population of over 100,000 persons. The time was ripe for it to be provided with more sophisticated public amenities such as a museum and a public library. The timing of this initiative, which was precipitated by the request of the Colonial Secretary, coincided with the crisis facing the hitherto privately operated Singapore Library and Museum.

On 24 January 1874 a notice appeared in *The Straits Times* announcing that a meeting of the subscribers of the Singapore Library would be held the next afternoon at the Exchange Rooms 'to consider what steps are to be taken to ensure better management for the future.' The announcement stated that the library 'has been in a state of bankruptcy for some time, and in order to work off the debt, the supply of new books has been stopped for the last six months.' The meeting would 'decide upon some means to pay off the debt and get out a new supply of books.'[34] We have no record of exactly what happened on the afternoon of 25 January 1874, but we do know that a decision was taken by the proprietors to ask the government to take over the library, pay off its debts and transform it into a public library.

Recounting the first thirty years of the library's life, Dr Robert Little told the Legislative Council that the government took over the library from its ten proprietors for $560.71 to pay the bookseller's bill then owing, with the promise that former subscribers continued to enjoy a lifetime privilege of using the library's facilities.[35] At the time it was taken over, the library had three sections, a lending library, a reference library and a reading library.[36] In all, the government acquired about 3,000 volumes which it immediately began cataloguing and repairing. It was opened to the public on 4 September 1874.[37] At the same time, the government tacitly assumed control of the Horticultural Gardens from the Agri-Horticultural Society, which had also run financially aground.

In 1868, after the Singapore Library moved out of the Singapore Institution, the latter was renamed Raffles Institution, in honour of its founder. Like the Singapore Library, Raffles Institution remained a private institution. It was only in 1903 that the government took over the school. On 20 March 1874, the Friends of the Raffles Institution held a meeting at the Town Hall over which Clarke presided. In the course of the meeting, Clarke informed the meeting that in view of Raffles' vision to

> …foster and promote the cause of education…some scheme would shortly be laid before the Legislative Council, by which the Institution should be made the means of affording instruction not only to children, but also to the adult population, and measures would be proposed for the establishment, in connection with it, of a Public Library and Natural History Museum, and in course of time, of elementary classes on literary and scientific subjects.[38]

The next day, at the opening of the Legislative Council, Clarke stated:

> The want of a public library, in connection with a museum, illustrating the products of the Malay Peninsula and Archipelago, is much felt here, and I am of opinion that such a library and museum might with advantage, be affiliated to the Raffles Institute.[39]

The Legislative Council acted swiftly and when it met on 28 March 1874, it resolved 'that a sum of $10,000 be voted towards the formation of a Library and Museum, on the understanding that the site, plans, and specifications, as well as the rules for its regulation, be hereafter submitted for approval to the Legislative Council.'[40] Dr Robert Little, who had been on the original Museum Committee, was appointed Chairman of the Library and Museum Committee. Explaining how the budget would be spent, Little told the Legislative Council that the Librarian would be paid $200 a month but that his principal duty 'would be as Collector and Curator of the Museum' and as such, one-third of his salary would be paid by the library and the remaining two-third by the museum.[41]

The man who was recruited as the first Librarian and Curator was the Scottish botanist James Collins, who 'appeared on the scene' as the government took over the library with references 'from some of the first men at home', including Sir Joseph Dalton Hooker, Director of the Royal Botanic Gardens at Kew.[42] Collins was engaged as Economic Botanist, Librarian and Secretary to the Committee and assumed duties on 8 May 1874.[43] Little envisaged that Collins, assisted by a collector, would go round 'and gather specimens of different textile fabrics, juices, and extracts and specimens of botany' and also 'make some expeditions' and purchase specimens for the museum. A budget of $600 was provided for the purchase of 12 wooden cases with plate-glass tops for exhibits.[44]

With the government takeover of the library and the engagement of Collins, the library appeared to flourish once more. At a committee meeting on 16 July 1874, Robert Little proposed that the library be named the 'Raffles Library.' This was rejected in favour of the 'Singapore Library and Museum'. However, this proposal was later withdrawn and the Committee approved the name of 'Raffles Library and Museum.'[45] In July 1874, the Chairman of the Library and Museum Committee petitioned the municipal commissioners for more space, complaining that the 'Library is cramped for want of extra accommodation.' The 'lumber room next to the Theatre' was allocated for its use.[46] The first batch of books ordered by the newly-reconstituted Library arrived in November 1874.[47] The library remained at the Town Hall till December 1876 when it moved back to Raffles Institution.

The Melbourne and Philadelphia Exhibitions

In February 1875, as the library and museum were finding their feet, a huge shot in the arm came by way of a letter from Sir Redmond Barry,[48] Chairman of the Victorian Commissioners for the Philadelphian Exhibition of 1876. The Centennial International Exhibition—to be held in Philadelphia from 10 May to 10 November 1876—was organised to celebrate the 100th anniversary of the signing of the American Declaration of Independence. It was to be the first official 'World's Fair.' As a prelude to British participation at this exhibition, a major colonial exhibition—the Melbourne Intercolonial Exhibition—was to be held in Australia in August 1875, and the Straits Governor was invited to forward 'such contributions to the Exhibition as will do justice to the Commerce and position of the Straits Settlements.'[49] The intention was 'to assemble at Melbourne

Museums in The Victorian Age

The modern museum is very much a product of the Victorian Age.[53] Wealthy collectors of curiosities and oddities have existed for centuries, but it was during the nineteenth century that the unique interplay of key political, social and cultural forces led to the widespread desire to have and to know the world in its entirety. This was an age that saw 'natural philosophers'—whose interests were 'very broad, ranging from astronomy to geology to natural history' and whose work consisted 'not only of laboratory experiments but of specimen collections and philosophical reflections'—supplanted by 'men of science,' who were much more specialised in their disciplinary endeavours and who saw their work 'as having broad applications and implications for society.'[54]

The collecting mania was driven by the need to study and understand nature and its infinite variety, and to educate through displays in the great temples of knowledge: libraries and museums. A major catalyst for the collection and gathering of scientific and natural objects were the big international fairs and exhibitions, the largest of which was the Great Exhibition in London of 1851. More museums had to be built to accommodate the scientific and technological instruments and natural products shown at these exhibitions. In London, it was thus no surprise to see the building of the Museum of Practical Geology in 1851, the Victoria and Albert Museum in 1852 and the British Museum of Natural History in 1881.

In this milieu, the colonial museums, like the Raffles Museum, served several purposes. First, it acted as a major collecting point for products, produce, flora and fauna of the colony to be shipped back to London for exhibition and study. Second, museums and botanical gardens provided the research needed for colonial servants to determine what kinds of crops might profitably be cultivated in the colony, and third, museums were places where the lay public could consume science and its wonders by wandering through its galleries and contemplating their exhibits.

1.8 The Great Exhibition of the Works of Industry of all Nations held at the Crystal Palace, Hyde Park, London, 1 May to 11 October 1851

all the objects which it is desired to be transmitted to America, and select from them those best suited for the purpose and to shop them conjointly to be placed on position at one and the same time in an Australasian Court.'[50]

The request was referred to the committee of the Raffles Library and Museum and Collins was asked to draft a list of specimen collections that could be assembled. Collins' list divided the specimens into six categories: (1) vegetable products; (2) animal products; (3) mineral products; (4) machinery and manufactures; (5) ethnology; and (6) general natural history collections such as insects, birds, shells, etc.[51] The *Straits Times* observed that this would be an excellent opportunity 'for swelling the collections' of the nascent museum.[52] It was also announced that the museum would publish a quarterly journal entitled *Journal of Eastern Asia* under the editorship of James Collins. The first issue of the first volume, released in July 1875, appears to be the only issue to have been published.

Space Constraints

Despite the fact that three of the best rooms of the Town Hall were reserved for the use of the library and museum, the space proved rather inadequate for the full realisation of both institutions. Collins was an excellent botanist but had no experience operating or curating a museum.

His immediate task after assuming office was to prepare a list of exhibits to be obtained for the Melbourne Intercolonial Exhibition. Collins requested that specimens to obtained in triplicate so that one set could be retained by the museum for local exhibition[55] Collins started off enthusiastically but soon became frustrated by the amount of time he had to spend on his library duties as well as the lack of space in the three rooms in the Town Hall. In his report for the year 1875, Collins complained about 'the total want of room'. Two showcases which he had arranged along the side of the reading room had to be removed and packed away when 'the space became necessary to the Library.'[56] This lack of space rendered it 'impossible to display any Museum objects' and Collins was 'forced to discontinue collecting' till proper space was found as specimens would quickly deteriorate in these poor conditions.

Notwithstanding these physical limitations, Collins came under increasing pressure to establish a proper museum. After all, money had already been allocated to the institution. In July 1875, a reader of the *Straits Observer* complained:

...The Museum itself is—well, if not a failure, it has not as yet advanced beyond being a collection of curios, such as an eccentric old bachelor at home with means at his command might have devoted the energies of his life to get together, to be dispersed at his death by the auctioneer at their full value and less than a sixteenth of their cost. Surely, the Curator could make more of it if he tried. His knowledge of botany is wasted on collecting specimens of shells and nasty looking snakes. The study of such 'objects' was not the object for which the public money was voted. The Museum was intended for a more practical purpose. It was to be maintained as a means of bringing together specimens or exhibits of the productive resources of the colony.[57]

The Straits Observer jumped to Collins' defence in September 1875 when it wrote:

We willingly grant that Mr Collins has so far as space at his command permitted him shown what a museum should be, but the want of room, and perhaps other annoyances, are great hindrances to him. If we are to have what some people think is a white elephant, let us make the best of the matter, and do what we can to render the animal useful. It is not acting sensibly to crowd library, reading room, and museum into three moderately sized dressing rooms. The present Secretariat would, if given up to Mr Collins, enable him to show what he can do; the present accommodation he has doesn't.[58]

In a review of the year 1875, the *Straits Observer* praised Collins for his work and argued for more space to enable him to establish a proper museum:

The Raffles Library is now under fair way and the Museum has during the last year received some few additions and was visited by over 4,000 persons. The Museum will, however, never be what it ought until it is located in a separate building, as it is impossible in the narrow limits within which it is now confined, for Mr Collins, the Curator, to arrange the many valuable specimens of all kinds, especially of produce, properly.[59]

The report ended by 'giving the greatest credit to Mr Collins for the earnest and able way' he performed his duties.[60] The space situation did not improve throughout most of 1876 since the plan to move the library and museum back to a new building in the Raffles Institution

Raffles Institution — Singapore.

1.9 Raffles Institution in the 1880s. The Raffles Library and Museum were housed in the three-storey block on the right between 1876 and 1887.

only materialised in December that year. In the meantime, Collins, who had ceased to actively acquire artefacts and specimens—having spent only $700 of the $3,000 allocated for acquisitions—, was still accepting donations. Since there was no longer any museum, as it were, he packed away the only two showcases.

At that point, things came to a head. The acerbic *Straits Times* asked: 'What must the King of Siam think when His Majesty hears that his "most valuable presents of Siamese commercial products, antiquities, and other objects" cannot be exhibited for want of space?'[61]

The Search for A New Home

Within two years of its establishment, it was clear that the library and museum were in need of a new home. The three rooms on the upper level of the Town Hall had very quickly become inadequate so much so that a large number of its artefacts and specimens could not be displayed at all. In May 1875, the governor asked the trustees of Raffles Institution to grant part of the unoccupied land behind the school as a site for the new Raffles Library and Museum building. The trustees agreed on the condition that they would be appointed ex-officio members of the Library and Museum Committee and that the new building would revert to the Institution if the library and museum should be abandoned.[62] Nothing much appears to have been done with the property after permission was granted to construct a new building on the site. In October, editors of the *Straits Times* urged the committee to get on with the building and suggested that two or three of its 'most energetic members' give the colonial engineer 'no peace until the Library and Museum Building is fairly started and finished.'[63]

The editorial of the *Straits Observer* noted in January 1876 that no work had begun on the proposed site and suggested that the new building which the school had built for the 'sons of Malay princes' be turned over to the committee for use by the library and museum since the school appeared to have no success in enticing the sons of Malay royalty to attend classes there.[64] In March 1876, at a sitting of the Legislative Council, Dr Robert Little asked the government why it had yet to commence work on the building even though $10,000 had been set aside for it some two years back. The colonial secretary replied that as the government was not able to spend the sum allocated, it would, as a temporary measure, 'hand over for the purpose the building now being erected for the accommodation of the sons of Native Rajahs in connexion with the Raffles Institution.'[65] The reason for not proceeding with the building was that another $27,000 would be required to complete it.[66]

The Raffles Library and Museum returned to its original home—albeit a different building—on the grounds of Raffles Institution on 26 December 1876. According to Collins, two floors of the three-storey new wing were placed at the disposal of the committee, giving him ample room for both a library and a museum. Although everyone assumed this move to be temporary and that funds would soon be found to build a new library and museum building, this would be the home of the library and museum for the next 11 years.

CHAPTER TWO

Laying the Foundations
1876–86

ALTHOUGH THE MOVE back to Raffles Institution signalled a new chapter in the life of the library and museum, especially that of the latter, James Collins was 'proving to be unstable and his behaviour had become erratic and unpredictable' by 1876.[1] He was dismissed but this left a vacuum in the administration. The museum needed to hire someone as a stop-gap measure.

A New Chapter:
Dennys and the Denizens (1877–81)

Dr Nicholas Belfield Dennys was born in 1839 and began his career in the Civil Department of the British Navy in 1855 aged just 16. In 1863, he joined the Consular Service in China where he was appointed Student Interpreter in Beijing. Three years later, he resigned and became Proprietor and Editor of the *China Mail* in Hong Kong where he was also Curator of the Hong Kong Museum from 1866 to 1876. As publisher and curator, Dennys was able to indulge in his love of Chinese subjects and published widely. His works include *The Treaty Ports of China and Japan: A Complete Guide to the Open Ports of Those Countries, Together with Peking, Yedo, Hong Kong and Macao* (1867), *Ch'o Hok Kai: A Handbook of the Canton Vernacular of the Chinese*

Language (1874), and *The Folk-Lore of China, and its Affinities with That of the Aryan and Semitic Races* (1876). He also edited *China Review* or *Notes and Queries on the Far East, Hong Kong* from 1872 to 1876. Sometime between his many duties, Dennys found time to obtain a doctorate and get called to the bar as well.

In May 1877, Dennys left journalism and Hong Kong for Singapore, not for a job at the library or the museum, but to take up his appointment as Assistant Protector of Chinese in Singapore.[2] This post had been created by a newly-passed Chinese Immigrants Ordinance,[3] and he was hired to assist William A Pickering, the first Protector of the Chinese in Singapore. Dennys was quickly made Justice of the Peace[4] and then served successively as third, second and first magistrate and commissioner of the Courts of Requests both in Singapore and Province Wellesley.

However, Dennys' experience as Curator of the Hong Kong Museum excited the committee into engaging him as a part-time Secretary, Librarian and Honorary Curator.[5] In August 1877, there appeared an announcement in the *Straits Times Overland Journal* informing the public that the General Committee of the Raffles Library and Museum and of the Gardens 'has been in a measure reorganized with a view to greater efficiency and authority.' The Colonial Secretary, John Douglas was chairman of the committee while Dr Robert Little was Vice-chairman of the gardens, and William Adamson was Vice-chairman of library and museum. Dennys was listed as Acting Librarian and Curator of the museum.[6]

Initially, Dennys spent two to three hours a week at the library and museum. His time commitment grew over time but he never accepted a permanent position there. We have no record of the artefacts and specimens Collins brought with him from the Town Hall to the new facility at Raffles Institution, other than those mentioned in the local newspapers, including the donation of the King of Siam. However, we do know that there were enough items in the natural history collection to keep the new curator busy for some time. In August 1877, Dennys told the local newspaper that he was 'not anxious to receive specimens of Natural History for a short period until the necessary cases etc. are completed, for their reception'. Dennys, who obviously leveraged on his Hong Kong connections, also obtained a 'large consignment' of specimens from the Hong Kong Museum.[7]

Despite only spending two hours a day at the library and museum, Dennys endeared himself to the establishment through his energy, hard work and welcoming disposition. By the middle of September, just a few months after Dennys took over, a journalist from the *Straits Times Overland Journal* visited the museum and praised him royally:

> We have much pleasure in testifying from personal inspection of the Raffles Museum, to the interest which Dr Dennys, the Secretary, displays in his labour of restoring the Museum or rather creating one out of an incongruous heap of specimens which hitherto lay in obscurity. Fourteen cases clean and attractive in appearance, have been filled with specimens of all kinds, some most valuable, and all more or less interesting. The improvements effected in the library itself strike a visitor prominently; a few coats of paint and whitewash have done wonders; and the careful classification of works under their various heads will render it an easy matter for the future reader who wishes to refer to any volume required.[8]

By this time, Dennys had the museum in order. In September 1877, the valuable geological collection had 'been placed in proper order' and Dennys was ready to receive new specimens.[9]

Beginnings of the Natural History Collection

One of Dennys' first efforts at securing local specimens was in August 1877 when the female rhinoceros at the Botanical Gardens zoo died. Alas, after the museum's taxidermist successfully extracted the skin from the carcass, it was found to be 'in very bad condition' and was 'judged unsuitable for stuffing.'[10] The skeleton, however, was 'very neatly mounted'[11] and put on display by the end of 1877. Dennys' call for donations met with a very positive response and one of his best supporters in this regard was the Temenggong Sri Maharaja, Abu Bakar (1833–95), son of Daing Ibrahim who initiated the museum with his gift of two Achinese gold coins back in 1844. Abu Bakar had succeeded to his father's title of Temenggong in February 1862 and took the tile of Maharaja of Johor in 1868. On 13 February 1886, the British recognised him as the Sultan of Johor.

In November 1877, Abu Bakar presented the museum with a live python. The intention was that the snake be killed and then stuffed and mounted. The same day, the snake was killed 'by the aid of chloroform and carbolic acid in the presence of the Maharajah and several leading residents.'[12] This particular python had a body circumference of some 33 inches and measured about 21 ft 2 in length. On the day the snake was killed, Abu Bakar's $1,000 racehorse, The Count, died. Charles Emmerson, a staunch supporter of the museum, managed to secure its skeleton for the museum.[13] Both the snake and horse were expertly mounted and put on display over the next two weeks.[14]

By November 1877, the insect cases were complete and on full display.[15] Also on display was a large Robber Crab that was claimed to be among the few preserved specimens of its type in museums. On 17 November 1877, Dennys issued a circular in which he invited members of the public to contribute specimens and artefacts. Dennys began by explaining that the museum's collections now comprised natural history, geology, economic botany, ethnology and 'export trade of Singapore' sections, and stated that the community could help by forwarding:

> (1)—Complete specimen collections of the produce etc, in which they do business, no dry package if possible to exceed four cubic inches in size, and each being carefully and accurately labelled with the *place of production, cost, name in English* and *Malay* and such other particulars as may seem advisable.
> (2) Specimens (labeled as above) of the different rocks, ores, earths etc found in or about mines with which they are connected.
> (3) Such models, more especially of any machines used in native manufactures, as their connection may render it easy to obtain: specimens of native fabrics, (in pieces of about 1 foot square) tools and implements used by native artisans etc.[16]

Dennys also appealed to Masters of vessels and others to contribute to the museum, and proudly informed the public that the 'Natural History department is now being effectively worked and every facility exists for preserving and mounting specimens of any size.'[17] The *Straits Times* commended Dennys' ability and energy and predicted that under his charge, the museum 'promises well to become, what it ought to be, an ornament that Singapore may well be proud of.'[18] In his Annual Report for the year 1877, Dennys stated that the considerable changes he had made amounted 'to a commencement *de nuovo*.' Despite his best efforts, the space allocated to the museum was still insufficient and even if the whole of the ground floor of the building were made available, it would but 'suffice for the next few years'. He expressed the hope that the government would set aside sufficient funds to build a 'Museum worthy of the singularly fortunate position of Singapore as a collecting centre.'[19]

In the course of the preceding six months, Dennys secured sufficient cases to display the specimens in the collection but could do nothing against the 'ferruginous dust so characteristic of Singapore' that permeated the building due to its exposed design. Furthermore,

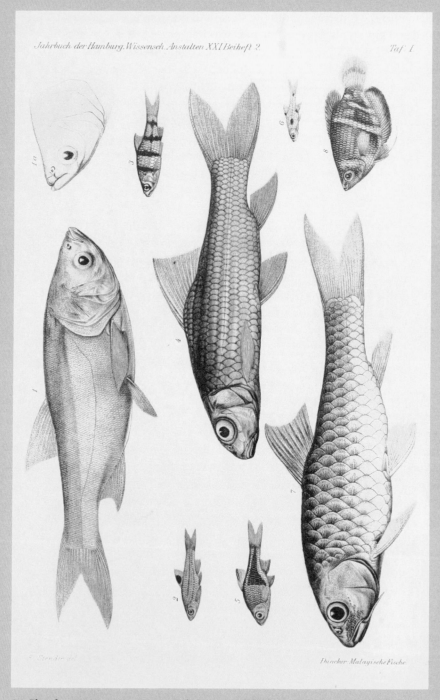

2.1 *Plate from an important 1904 paper by German ichthyologist Georg Duncker, who described many new species from the Malay Peninsula in the 1900s. The characteristic Harlequin Rasbora (Trigonostigma heteromorpha) (Fig. 5, last row, second from right) was described from specimens he collected from the Botanic Gardens in September 1900 when he was hosted by R. Hanitsch.*

2.2 *Plate of scallops from George Brettingham Sowerby II's* Thesaurus Conchyliorum *in which many new species were first named. One of the most famous conchologists of his time, Sowerby described many new shell species from Asia, including the appropriately named Singapore Scallop* (Volachlamys singaporina) *(Fig. 55, second row from bottom, third from left).*

not everything could be put under glass, especially the larger exhibits. Dennys noted:

> Such specimens as it is impossible to put under-glass (the rhinoceros skeleton, and python, for instance) are daily found to be coated with a fine red dust which materially deteriorates from their appearance despite every effort to the contrary.[20]

Dennys then proceeded to give the most thorough report on the state of the museum as at the end of 1877. As this is the first comprehensive survey of the collection since its founding, it is worth reproducing Dennys' survey as follows:

> **Geology**. A perfect collection obtained from Professor Tennant in England has been properly arranged, and the commencement of an interesting local series has been also made. The Catalogue of this department is ready for printing.
>
> **Economic Botany**. A fair collection, purchased from Mr James Collins, has been arranged, to suit general convenience, in alphabetical order, and the Catalogue is ready for printing. Contributions of local specimens are earnestly desired and invited.
>
> **Conchology**. A small collection of shells has been made, and is receiving daily additions. This department owes much to Mr C Emmerson and it is hoped to shortly commence classifying and cataloguing the specimens.
>
> **Entomology**. A small but very perfect collection of Sarawak Lepidoptera fills four of the cases, and local specimens of the various orders are being gradually accumulated. Mr We B Pryer (late Hon Curator of the Shanghai Museum) has lent me material assistance towards the formation of a Catalogue.
>
> **Ornithology**. The cases now contain 63 mounted specimens... Acting on the suggestion of a distinguished Indian Ornithologist, all birds received which can positively be pronounced natives of Singapore are retained unmounted, awaiting his visit to this part in the spring, when he will aid in their accurate identification. Two small collections, one of Borneo birds and one of Malacca birds, have been sent to the great Ornithological Museum in Simla for a similar purpose, while the Curator of that Institution has despatched a large and handsome collection of Indian birds to this Museum free of cost.
>
> Of **Mammals**, the Museum as yet can show scarcely any mounted specimens...But the Museum possesses three skeletons of value, that of a Sumatran rhinoceros, horse and tiger besides sundry skulls etc.

Of **Ophidia**, a very fair but as yet unidentified collection, both local and exterior, is in place, prominent amongst it being the skin and skeleton of a very fine female python, presented by HH the Maharajah of Johore. The latter forms a really beautiful preparation and reflects great credit on the Museum Osteologist. It is noteworthy that of the numerous specimens of the serpent tribe presented since I have been in charge, only on *Cobra de Capello* (and that a very small one) has come to hand. There is an ample field for research respecting the Ophidia and Phanatophidia of the island and peninsula, in which I hope some day to labour. Meanwhile residents can effectively aid by sending every specimen caught to the Museum with full details as to their locality. Native name etc. Of **Chelonia** there are 3 specimens in the Museum.
Of **Ethnological** specimens the Museum possesses a very fair share, as a beginning and as such objects are always attractive, their presentation is highly valued.

Dennys concluded by calling upon the public to continue supporting and contributing to the museum and informing the committee that he has been in contact with 'many of the leading men at home and elsewhere in the various branches of Natural History' in an effort to 'aid the Institution to make it of high scientific and popular interest.'[21]

It is important to note that with Dennys' reorganisation, the seeds of a natural history section within the museum were sown. More than that, Dennys systematically organised the collections to give pride of place to natural history and made a distinction between mounted specimens for show, and scientific reference specimens that were not mounted, but kept for future study by scholars and scientists. In doing so, Dennys was following in the footsteps of John Edward Gray, Keeper of Zoology at the British Museum who strongly advocated the need to keep reference and display collections separate.[22]

Some Interesting Acquisitions

In March 1878, Abu Bakar, obviously pleased with how his donations of a python and a racehorse had been displayed, decided to present the museum with 'a fine live tiger' that the acting curator was to 'take steps to kill and preserve him for the institution.'[23] Dennys proceeded to the Maharaja's Istana Besar in Johor Baru and killed the tiger with a huge dose of strychnine. The poor animal died about four hours after ingesting the first dose of the toxin.[24] Another famous supporter of the museum was Whampoa Hoo Ah Kay, one of Singapore's leading

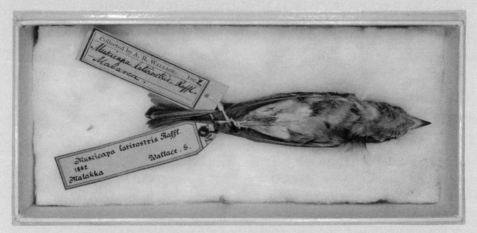

2.3 *The only Wallace specimen in the museum. The stuffed bird, an Asian Brown Flycatcher* (Muscicapa dauurica), *collected from Malacca, was apparently passed to Alfred Wallace by a German collector; identified by him; and eventually passed to the museum as part of an exchange with the Berlin Museum.*

2.4 *Specimens collected by Wallace from Southeast Asia in his original cabinets, now at the Natural History Museum in London*

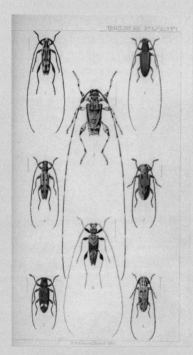

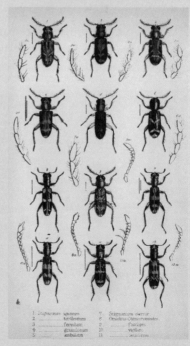

2.5 *Longhorn Beetles named by Francis Polkinghorne Pascoe in 1864, including Eoporis elegans (Fig. 6), collected by Wallace in Singapore*

2.6 *Beetle plate from Obadiah Westwood's 1855 paper. Westwood studied Wallace's collection and every species here, all from Singapore, was new to science.*

Chinese businessmen and community leaders. Hoo was one of the original founders of the Agri-Horticultural Society that later became the Botanic Gardens. When his pet orangutan died on 30 June 1878, he donated it to the museum for mounting. The Botanic Gardens Zoo, which functioned between 1875 and 1905, was also a source of donations when the animals died (*see* Box Story, p.32). Ordinary folk, when the opportunity presented itself, also sent donations to the museum. For example, a Malay fisherman captured a 12-foot long crocodile in Sunghie Battu Belyhar near Telok Blangah in July 1878 and brought it to the Central Police Station, which proceeded to send it to the museum.[25] Another crocodile, this one measuring 11 ft in length, was killed and sent there by John Fraser of Trafalgar Estate after it had devoured a woman on the Punggol River in 1886.[26]

The following items were presented to the museum after the animals were hunted: a large male peacock shot by a group of young men in July 1879;[27] the head and antlers of 'a splendid deer—a buck—weighing 500 lbs' shot by a Mr Barclay in August 1879; an old tiger, said to be 50 years old, shot by an old Malay plantation caretaker in Telok Paku in November 1878;[28] the skin of another large tiger, measuring 8 ft 9 in length, which was shot near Tampines in January 1886 by Daniel Maw; and lastly, a panther, shot at Carrington House on Mount Sophia.[29]

One of the most remarkable species collected by the museum was the *luth* or Leathery Turtle (*Dermochelys coriacea*). A dead specimen was found on the beach near Siglap in 1883. The Malay villagers were

2.7 Leathery Turtle found dead on the beach near Siglap in 1883 by Malay villagers

trying to throw it back into the sea when A. M. Skinner, the Inspector of Schools, chanced upon them and persuaded them to bring the turtle into town.[30] Measuring some 6 ft in length, it is now one of the treasures of the Zoological Reference Collection. Other unusual and rare natural history specimens include a very large blood python (*Python brongersmai*), brought to the museum by the police in Johore in April 1883,[31] and a long-legged Japanese Spider Crab (*Macrocheira kaempferi*)[32] measuring 9 ft 6, obtained for the museum by one Edgar Abbott of Yokohama through local resident George Mildmay Dare.[33]

Dennys appears to have forged an excellent working relationship with Allan Octavian Hume's private ornithological museum in Simla as it generously donated many bird specimens to the Raffles Museum. Hume's curator and collector from Simla, William Ruxton Davison, also visited Singapore regularly to study the collection and in the process, also added to the collection.[34] In 1881, Dennys reported progress

> …in the collection of mammals, birds, and insects, and some valuable additions have been made to the herbarium. It is, however, in reptiles that the most satisfactory and practical results have been attained. All the specimens have been identified by high scientific authorities at home, and before long the Museum will possess a complete and unique collection of Malayan Reptilia. Two species of serpent, new to science, have been discovered, and other additions may be expected when the identification of the latest consignment sent is completed.[35]

The Raffles Societies Ordinance

It may be recalled that in 1874, the Straits Government took control of the library and museum as well as the Agri-Horticultural Gardens by an outright purchase of its debts, and by assuming control over its properties and chattels. The administration of these three institutions was nominally in the charge of the governor who appointed the colonial secretary as chairman of the Raffles Library, Gardens, and Museum Committee. The legal status of these three institutions was, however, unclear, and the Colonial Office in London required the local authorities to clarify the ownership status of the properties associated with these bodies. The colonial secretary, John Douglas, penned a long memorandum detailing the history of the transactions involving the three institutions. He then informed the colonial office that the attorney-general, Sir Thomas Braddell, was of the opinion that in respect of a large tract of the gardens, the committee held the land on

The Zoo at the Botanical Gardens[36]

The origins of the Singapore Botanic Gardens can be traced to 1859 when the Agri-Horticultural Society was granted 32 hectares of land in the Tanglin district by the colonial government. The pleasure gardens were laid out by Lawrence Niven. In 1874, the society ran out of funds and the government took over the management of the gardens. Shortly after, it was decided that a zoological garden be established within its premises. Once this was made known, the gardens received numerous valuable donations of live animals. One of the first to arrive was a Sumatran Rhinoceros, which had been presented by Datuk Klana of Sungi Ujong to Governor Sir Andrew Clarke. A large 'house' was built for the creature as the management hoped to include an elephant and some tapirs to the collection. Within its first year of operations, the gardens received a Sumatran rhinoceros, a sloth bear, two tiger cats, two sambur deer, a civet cat, a great kangaroo, two red kangaroos, a bush-tailed wallaby, four black swans, three Australian rabbits, two orang-utans, and a whole host of birds and other smaller animals.

In 1876, gardens superintendent Henry James Murton reported that he received a 'fine young tiger' from the Sultan of Terengganu; a leopard from a Siamese minister; and a leopard from W. Hargreaves. Unfortunately, both these leopards were poisoned by a hooligan and died. They were donated to the museum. The Sumatran Rhinoceros came to a rather sad end when the committee grew 'tired' of the creature and decided to dispose of it. We know not how it died but as noted above, its carcass was donated to the museum for mounting. Unfortunately, the skin was in poor condition and could not be mounted. As the larger creatures died, the museum received their skins and carcasses for mounting.

At its height, the zoo at the Botanic Gardens had some 140 animals. However, 'without sufficient funds to maintain them, gradually by death, misfortune, sale or gift, the animal population faded to a few birds and monkeys.'[37] In 1905, the zoo—which was often referred to as 'The Menagerie'—was officially shut down and the surviving animals were donated to the London Zoo.

2.8 *Mounted display of young leopards, circa 1904*

trust for the President and Treasurer of the Agri-Horticultural Society!

Sir Michael Edward Hicks-Beach, Secretary of State for the Colonies was concerned that with the Straits Government contributing a large portion of the budget to maintain these three institutions, 'it is necessary that Government should exercise a more direct control over their management than was contemplated in the draft Ordinance prepared in 1875'. Hicks-Beach suggested that in such a situation, the committee should 'become merely a consultative body, and should be appointed wholly by the Government'.[38] To this end, a new draft Ordinance was prepared and this was passed in the Legislative Council as the Raffles Societies Ordinance on 16 December 1878.[39]

The Ordinance had a long preamble, but its short title succinctly states the object of the legislation: 'To confirm arrangements relating to the Agri-horticultural Society and the Raffles Museum and Library'. Section 5 of the Ordinance put the matter of property ownership beyond doubt by making the three institutions and 'all growing and dead plants, trees and flowers and other property in the said gardens, and all books, papers and manuscripts, and other property in the said Library, and all articles in the said Museum shall be deemed to be public property'. Under section 2, the Botanical Gardens of Singapore and the Raffles Library and Museum 'shall be managed as Government establishments by a committee or committees to be appointed by the Governor for the purpose'. Thus, from a legal perspective, the Raffles Library and Museum became public institutions with effect from 16 December 1878.

A Home of Its Own

While the government was sorting out the legalities relating to the library, museum and gardens, the old question of space reared its inconvenient head once more. In March 1878, it was reported that Raffles Institution now needed the portion of the school that had been given up for the library and museum's use. By this time, the museum had become a major local attraction. In May 1878, it was reported that between 1,500 and 2,000 visitors thronged the museum daily, and most of them were 'Natives of all classes'.[40]

The museum became so crowded that rules were imposed to stagger the crowd. The patently racist *Straits Times* report of 25 May 1878 states:

> The daily influx of native visitors to the Raffles Museum having reached an extent which almost precludes ladies from entering

it, the rule has been adopted of reserving from 10 am to 1.30 pm for natives, coolies etc and from 2.00 pm to 5.00 pm for European and other respectable visitors and ladies. The intervening half hour is devoted to sweeping and washing the upper floor, as it rapidly becomes covered with mud and dust during the forenoon.[41]

Two months later, another newspaper report suggested that a small entrance fee of five or ten cents be charged, and urged that:

As a great number of presentations are made every month to the Museum by residents and strangers, a separate building for it and the Library, as originally intended, will soon become an urgent necessity.[42]

Dennys was clearly pleased with the popularity of the museum and with the tremendous support he was receiving from the committee and from the public generally. In his report for the year 1878, his only lament was that while he was almost deluged by the presentation of specimens by well-wishers, he received 'no scientific aid' from residents, except with respect to botany.[43] By now, the museum had clearly overtaken the library in terms of popularity and physical footprint. Commenting on Dennys' 1878 report, the *Straits Times* wrote that there was 'too much attention' being paid to the museum, 'much to the detriment of the Library',[44] which did not enjoy a good supply of new books. The article went on to champion the need for a new building for the museum, reminding readers that in 1874, a budget had already been set aside for 'a separate and special building for the Raffles Library and Museum', and urged the government to make good on their promise.[45] While it was true that a vote had been approved for the erection of a new building for the library and museum, the finance committee reluctantly agreed 'to recommend that the proposed vote of $13,000 for the erection of a Library and Museum be struck out of the Estimates, as it appears to them that a Lighthouse on *Pulo Pinang* is a work of more pressing necessity.'[46]

In the meantime, the pressure on space became ever more pressing. Raffles Institution, which had large increases in its enrolment, was desperate for more classroom space and was constantly egging the library and museum to move out. However, with the legal transformation of the library and museum into public institutions, this became increasingly difficult for the school.[47] At the same time, Dennys was also out of space given the numbers of donations and acquisitions. Indeed, in his report for 1880, Dennys wrote:

> Owing to the restricted amount of room at command and the
> hopes entertained of before long obtaining proper accommodation,
> the Committee have judged it unwise to expand any large sums on
> purchases for the Museum. It was thought better to treat the sum
> shewn as balance credit as a reserve to meet a portion of the heavy
> expenses which the fitting up of the new building will entail.[48]

In March 1883, the Acting Colonial Secretary announced that the
governor had received a dispatch from the Secretary of State, approving
the estimates of the current year save for that of the building of the
museum.[49] This was a major blow to both the committee as well as the
Trustees of Raffles Institution who were getting desperate due to the
space shortage.[50] In October 1883, the Straits Government once again
included a budget for the construction of the library and museum. A
sum of $20,000 was included and this time, it was approved.[51] A site at
the slope of Fort Canning Hill was identified and chosen for the
building. Sir Frederick Weld, governor of the Straits Settlements proudly
announced this at a meeting at the Royal Colonial Institute in London:

> A museum, which is proposed to build on a large scale, which will
> render it the most complete institution of the kind in that part of
> the world, is to contain a department for industrial exhibits and a
> library. I am anxious to establish a scientific department in charge of
> it. Our mineralogical resources are little known. Our territories and
> neighbouring islands and countries contains numberless products
> which may be made useful. In botany, zoology, ichthyology,
> entomology, much remains to be done, and not only the colony but
> science generally will be benefitted.[52]

The colonial engineer, Major Henry McCallum, was tasked with
the design of the building since the Colonial Office thought it
'hardly worthwhile to submit them to a first rate English architect.'[53]
McCallum's first plans were ambitious and grandiose and came with a
tag of $130,000. They were rejected on account of cost and MacCullum
was asked to scale down the design and keep within an $80,000 budget,
which he did. The plans were submitted in May 1882 and in October
1884, contractor Chan Ah Quan was engaged to construct the building
for $61,000. Unfortunately Chan did not complete the building.
Gretchen Liu tells us that the construction of the dome proved so
problematic for Chan that he went mad and gave up.[54] Construction
was assumed by one See Ah Tock who finished the building.

It was only in October 1884 that the public first came to know of
the exact plans for the library and museum:

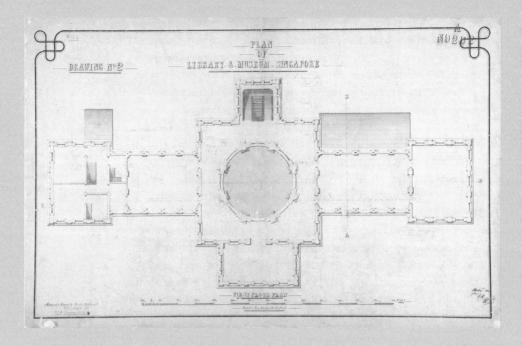

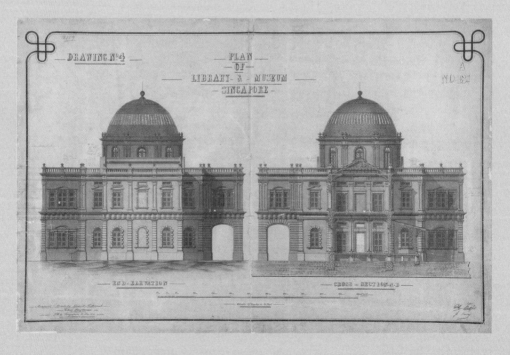

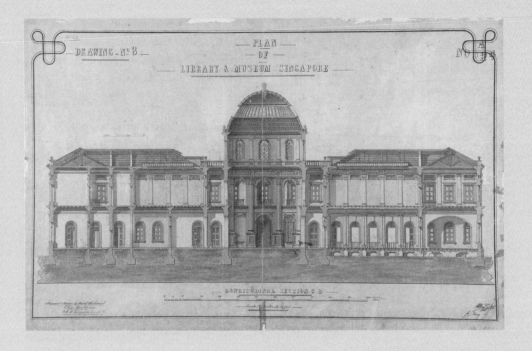

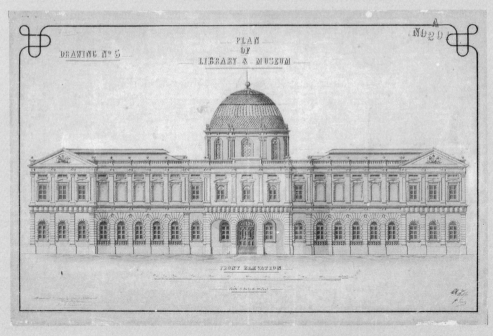

2.9 *Construction drawings of the proposed Library and Museum at Stamford Road. The building was designed by Colonial Engineer Major Henry McCallum and was built between 1882 and 1887.*

> The building is to be two stories high, having an imposing dome with variegated lights in the centre, and ornamental eaves all round. Under the dome will be a well, surrounded by a circular gallery, and in the wings of the upper floor will be stationed, in separate apartments the natural history, mineralogical, ornithologicial, and other departments of the Museum, with plenty of room for expansion in each department. The Library will be located on the ground floor, as will also be the Curator, who will at first have quarters near his department.[55]

Work began in earnest in December 1884 with bricklayers building a 'dwarf-wall' and 'cutting away the big blocks of stone' on the site,[56] and by January, the foundation of the building was ready.[57] As the building was being constructed, Arthur Knight, the acting curator, was completing the first catalogue of the museum. The project had begun in 1882 at the request of the committee and was completed and published in 1884. It presented 'a complete list of the specimens illustrating Natural History, Botany, and Mineralogy' save for specimens that could not be identified.[58] The collections continued to grow through exchanges with other museums. For instance, the Raffles Museum received 61 birds and 17 other animals from the Australian Museum, and 155 birds from the Queensland Museum. The latter was sent in exchange for surplus specimens sent over by the Raffles Museum.[59]

In January 1886, the government announced that it had set aside a large room in the new museum for a reference library and a 'home' for the Straits Branch of the Royal Asiatic Society, which had been established in 1877.[60] By August 1887, the building was close to completion. On 11 August 1887, a notice appeared in the local papers announcing the closure of the library and museum in the month of September 'to allow of the books and specimens being removed to the New Building.'[61] The notice was signed by George Copley as Acting Secretary and Curator.

The magnificent library and museum building was declared open on 12 October 1887 by Governor Weld, who was officiating in one of his final duties before retiring. The opening took place at 5.00 pm amidst much pomp, speeches and ceremony. The Library and Museum had found a place it could all its own—at last. A hundred years later, Singapore would celebrate the museum's 'centenary' with equal pomp and circumstance.

Leadership Transitions

In 1881, Dennys went on a four-month home leave. The same year, proving himself once again as a multi-faceted talent, Dennys even patented a chemical coating for ships' bottoms and ironwork exposed to immersion in seawater.[62] His position at the museum was taken over by Arthur Knight for $50 a month. Knight, who arrived in Singapore in 1860 to assist his brother-in-law George Henry Brown (after whom Bukit Brown Cemetery is named) in business, continued to act, on and off, as Curator till Knight's death in 1916.[63] In June 1882, Dennys requested that the position of Curator be made separate from that of Librarian. Arthur Y. Gahagan was thus appointed Acting Secretary and was in charge of the library while Dennys retained the Curator's post. Dennys functioned as Acting Secretary till May 1882 after which he ceased active curatorial work while remaining on the committee at least till 1886.[64] Dennys would be 'invalidated home in 1889' and would resign from service in May 1889. In June 1899, he took up the editorship of the *Manila Times* but had to abandon the post on account of poor health. He would die in a Hong Kong hospital on 5 December 1900 following an operation.

After 1882, Dennys's curatorial duties devolved to Knight and his library duties to Gahagan. The Committee discussed the appointment of a suitable Curator and Librarian in March 1886 and George Copley was appointed. In mid-August 1887, Copley left to take up his post as Secretary to the Municipality in Malacca and his position was filled by W. T. Wrench.

Realising that the library and museum could no longer rely on part-time secretaries and curators, the committee had, as early as July 1886 approached the British Museum for help in finding a 'Curator and Librarian for the Raffles Library and Museum, who will also be the Secretary to the Committee of Management.'[65] In January 1887, the committee received an application from one Dr Robert von Lendenfeld.[66] According to Hanitsch, the committee agreed to engage Dr von Lendenfeld but his agreement was cancelled at the last moment for some unknown reason.[67] It is possible that von Lendenfeld himself had chosen not to come to Singapore, having secured a position of assistant lectureship at University College, London. In May 1887, the committee decided to hire William Ruxton Davison the ornithologist from Simla. Davison officially joined the Raffles Library and Museum on 6 December 1887.[68]

CHAPTER THREE

The Halcyon Years
1887–1926

THE OPENING OF the new building and the engagement of William Ruxton Davison as full-time Curator augured well for the institution. With a building it could finally call its own, the Raffles Library and Museum had now the space to properly display its ever-increasing collection, and adequately provide for library patrons as well.

The Davison Years: 1887–93

3.1 William Ruxton Davison, the museum's Curator from 1887 to 1893, also an expert bird collector

Davison was a most promising hire. Prior to taking up his position at the Raffles Museum, he had been Collector and Curator for the great British bird collector Allan Octavian Hume (1829–1912). Hume, who has been called the 'Father of Indian Ornithology' was a British civil servant who, among other things, co-founded the Indian National Congress.[1] Davison was known as one of the best field naturalists of his era,[2] and having a command of Malay, Hindi and Tamil, much was expected of him as Curator.[3]

While previous curators of the museum spent much time and energy fighting the red lateritic dust that coated exhibits every day, Davison had a different problem—mould. The conditions in the museum were attributed to the rising damp in the walls and the impossibility 'of keeping out the same on rainy or cloudy days.'[4] This problem was to some extent alleviated with the installation of glass windows but these made the building's interiors rather dark. By the end of 1888, Davison was able to report that the 'difficulty experienced in 1888 of keeping the Zoological collections, specially the insects, free from mould, has not been so great during the past year, and as the walls of the Museum become thoroughly dry this difficulty will be considerably lessened.'[5]

While the problem of damp seemed to have improved with passage of time, this was not the end of Davison's problems. In September 1891, the museum would suffer a major termite infestation, which would only be the beginning of a long-term problem. The termites attacked the staircase but fortunately left all other exhibits intact.[6] Some issues of newspapers and 'many copies of the *Pall Mall Gazette*, and all the remaining copies of the 1st Supplement to the Catalogue' were destroyed. Davison's curator's quarters were severely attacked, forcing him to vacate,[7] and his personal collection of rare butterflies was damaged.

The high expectations of the museum were further heightened with some notable specimen contributions in 1888. The first was a specimen of a young Malayan tapir presented by Governor Sir Cecil Clementi Smith. The specimen had been carefully mounted and was to be placed alongside the adult tapir specimen already in the museum's possession. The second was an enormous crocodile (15 ft 5) shot and donated by George Paddison Owen, one of Singapore's most well-known big-game hunters and colourful characters.[8] The Assistant Superintendent of Police, Rowland Conolly, presented the museum with a 21-foot reticulated python, and two sawfish came by way of Dr Thomas Rowell (Principal Civil Medical Officer) and Captain J. G. Trewin of the *SS Meanatchy*.[9] In 1889, 'high profile donors' included George Copley—former Curator of the museum and Secretary of the Muncipality in Malacca—who purchased and donated 'a nearly adult tigress, an adult female Sun Bear (*Helarctos malayanus*), and an adult male of Temminck's Golden Cat (*Felis temmincki*).' John Pickersgill Rodger, British Resident of Pahang also presented the museum with a 'fine adult cow of the seladang (*Bos gaurus*), that year.[10]

Popularity and Extended Hours

The museum continued to be extremely popular among the locals. As Davison informs us in his annual report for 1888, a large number of visitors requested him to open the museum on the same nights as the reading room.[11] Davison considered this proposal quite seriously but concluded that while it may 'double the number of visitors' to the museum if it opened at night, it would be necessary to light it with 'at least 20 good lamps' and such kerosene lamps would pose a major fire hazard.[12] Besides, the 'windows would have to be kept closed (for the damp night air would soon ruin those specimens not under glass)

3.2 *The enormous 15 ¹/₂ ft crocodile shot by George P. Owen in 1887 at the Serangoon River and presented to the museum*

the large amount of heat developed would make the rooms very hot, and if the number of visitors was large, the air would soon become unfit to breathe.' Of course, all these potential problems would be easily overcome 'if the building were lit by electricity.'[13] Davison's last statement was really more of a pipe dream than a realistic option at the time, since the first private electric power station was only built in 1906.[14]

Davison was so encouraged by the large attendance at the museum that in 1889 he reiterated his suggestion to have it lit by electricity since there were several occasions when 'parties, chiefly composed of women and children, Chinese or Malay, have come to the building on the night the Library was open, or to [his] quarters where they saw the lights, asking that they might be allowed to visit the Museum.'[15] If this came to pass, he proposed opening the museum from 7.30 pm to 9.30 pm on week nights. He further proposed opening the museum on Sundays from 8.00 am to 11.00 am and then again from 1.00 pm to 4.00 pm as these hours 'would not interfere with the services at the Cathedral.' He was certain that this move would also be 'appreciated by many connected with the shipping, to whom Sunday is the only off day.'[16]

Attendance at the museum continued to increase between 1889 and 1891, but 1892 saw a mysterious decrease in attendance—a drop of 14,147. Davison stated that this 'great falling off in numbers' could not be satisfactorily explained.[17]

Reorganisation and Layout of the Museum

The year 1889 was a hectic one for the museum. According to Davison, the collections—especially the zoology collection—increased 'so rapidly' that he could not 'get specimens mounted and exhibited as rapidly' as he wished.[18] The work of the Raffles Library and Museum kept Davison so busy that he had no time to rearrange the exhibits in the museum, something he had been particularly anxious to do.[19] Nonetheless, Davison was able to hire two 'native' Collectors in April 1889 in lieu of the Fish Taxidermist, and they were sent off to collect specimens in Selangor, from Kuala Lumpur to the borders of the state of Pahang. They also collected 215 birds, 12 small mammals, 30 reptiles, and some hundreds of insects in Singapore.[20] At the end of the year, Davison reported an increase in expenditure over that of 1888; this was due to the 'extensive fittings required in the Library, Museum and Workshop', an increase in salaries (of some of the older employees and of Davison), 'and the publishing of a new catalogue.'[21]

In his 1892 report, Davison once again raised the matter of rearranging the museum's entire collection. He noted that a large number of ethnological specimens were accumulating but there was no room for exhibiting them. He even suggested that the former Curator's quarters—which he recently vacated on account of the termite infestation—might be usefully added to the museum's display space and provide room for a large ethnological collection.[22]

Defining a Collection Strategy

Up to this time, the museum had been content to collect practically anything and everything that came its way. Previous curators have been happy to accept donations of all sorts and in the preceding decades, accumulated a large assortment of specimens. Where the museum expanded its own funds in acquiring specimens, it focused on specimens from Singapore and the Malay Peninsula but there was no collection protocol to speak of. One of the most important decisions Davison made was to announce a decision to limit the collection of specimens to the Malay Peninsula and 'adjacent countries'. In his 1889 Annual Report, Davison contemplated the numerous Australian specimens in the museum's collection, and opined that:

> It would be advisable, I think, to confine the contents of the Museum to the products of the Malay Peninsula and the adjacent countries which are zoologically affined to it, and this is best indicated by what may be termed Wallace's 50-fathom line, which includes only Sumatra, Java, Bali and Borneo, to the South. As specimens from any of these localities are obtained they will replace the Australian and other specimens from places beyond the shallow sea line. One exception to this should be made with regard to the Cocos-Keeling Islands, which lie well out of the 100-fathom sea, and which, as might be expected, show an affinity with the Australian fauna; but being connected with the Government of the Straits Settlements, it would be interesting to get as complete a collection from there as possible.[23]

The 50-fathom line referred to by Davison was set out by the great naturalist Alfred Wallace in his description of the physical geography of the Malay Archipelago:

> Returning now to the Malay Archipelago, we see that the whole of the seas connecting Java, Sumatra, and Borneo with Malacca and Siam are under 50 fathoms deep, so that an elevation of only 300 feet would add this immense district to the Asiatic continent. The 100 fathom line will also include the Philippine Islands and the island of Bali, east of Java.

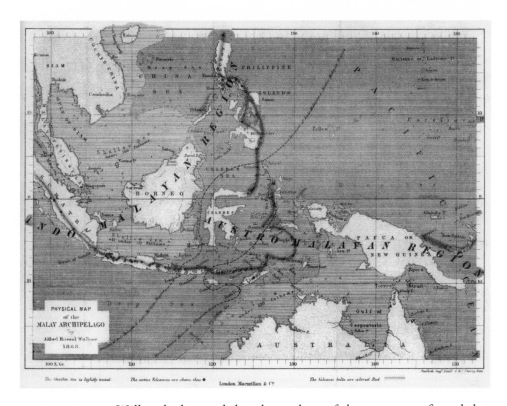

3.3 The Wallace Line, a faunal boundary line delineated by Wallace in 1859, from The Malay Archipelago (1869)

Wallace had argued that the zoology of these areas confirmed that 'these great islands must have once formed a part of the continent, and could only have been separated at a very recent geological epoch.' This leads us to the irresistible 'conclusion that at a very recent geological epoch the continent of Asia extended far beyond its present limits in a south-easterly direction, including the islands of Java, Sumatra, and Borneo, and probably reaching as far as the present 100 fathom line of soundings.'[24]

Davison took steps to effect this collection policy. In 1889, he received donations from Dr George Darby Haviland of Sarawak of a 'fine female and young of the Bornean Orang-Utan'; a fine mounted series of adult and young hoary bamboo rats (from Malacca); a 'fine collection of mammals and birds collected at Baram in Borneo'; mammals and birds collected by the museum's collectors in Singapore and Johor, including a rare civet cat known as Hardwick's Hemegale; a large moth collection from Colonel Charles Swinhoe; and a miscellany of specimens from the eastern part of the Malay Peninsula, and Malacca.[25]

Davison himself was involved in several collecting expeditions, the longest being a two-month expedition to Pahang commencing on 23 June 1891. The expedition, which was led by Henry Nicholas

Ridley, Director of Gardens and Forests of the Straits Settlements, also included Lieutenant Kelsall of the Royal Artillery.[26] The expedition was postponed to June most probably due to the death of Davison's wife on 27 March 1891 from bronchitis[27]. They returned to Singapore on 9 September 1891, having been away for two months and thirteen days. The expedition yielded 'a good series of the smaller mammals— monkeys, squirrels, etc.' though none of them were 'new to science among them.'[28] These were, in due course, mounted.

Davison's Suicide

While the challenges at the library and museum were numerous, Davison seemed to manage the situation adequately. However, his wife's death drove him into a state of depression and to drinking. Towards the end of 1892, his alcoholism and mental condition worsened. He was admitted to hospital in December 1892 and Dr Ellis, who examined him on 7 January 1893, considered him 'to be of unsound mind' and stated that 'he was not responsible for his actions.'[35] He was suffering from hallucinations and told Dr Ellis that 'he intended to commit suicide.'[36]

On 21 January, Davison discharged himself and went to stay at the Victoria Hotel. On the evening of 24 January, Davison even wrote out a memorandum authorising the Library Clerk, Chua Hood Leng 'to act for me and in my behalf entirely.'[37] In recent years, Davison had come to depend on Chua, whom everyone considered to be extremely capable, conscientious and reliable. Alas, Chua would come to grief a few months after Davison's death when he was charged with criminal breach of trust.[38]

On 25 January 1893, Davison was found dead in his room at the Victoria Hotel. A Coroner's inquiry ensued and it was discovered that he had 'died from the effects of an overdose of opium taken whilst in a state of unsound mind.'[39] Quite amazingly, Davison still managed—in this poor condition—to submit the 1892 Annual Report, which was dated 14 Jan 1893. It would be published later that year.

Davison's death brought to an end the first significant period of the museum's growth and development. Contrary to Liu's suggestion that his tenure was 'fraught with troubles,'[40] he seemed to have successfully increased the museum's popularity and put the collection on a proper, scientific footing. His determination that the collection of specimens be based on Wallace's 50-fathom line was perhaps the most significant taxonomical imperative that the museum had developed to date. He also made an important start in the organisation of regular collecting

The Whale

In 1892, the museum acquired its most iconic and unforgettable specimen and exhibit, a 42-foot skeleton of an Indian Fin Whale (*Baleaenoptera indica*). The whale was stranded at Sa'Batu near Malacca on 17 June 1892. News of this stranding first reached the shores of Singapore on the 25 June when the British steamer *Bengkalis*, reported seeing a whale 'stranded on the beach near the Kesang River'. Captain Angus 'went out in a boat and examined the monster, apparently a male which had been dead some few days, and was even then in a highly savoury condition.'[29] It was reported the whale had died two days earlier, on 23 June. Another contemporary account reported that boatmen from Malacca brought word that 'a fish of monstrous size (*ikan besar skali*) had been stranded upon the shore of the Kessang River.'[30] The day before, a party from Muar, including the Resident, DPA Harvey, Ungku Sulieman, and other chiefs proceeded to check out the find. According to a contemporary account:

> On nearing the spot, somewhere about Sebatu in Malacca the carcase came in sight at a distance. The party found much difficulty in getting near to it owing to the depth of mud around. They found that it was the carcase of a whale. Its extreme length was 44 feet, unfortunately it was on its back and the upper surface of the head was imbedded in the soft mud so that no observations could be made as to the existence of a dorsal fin etc—it bulked largely being about 9 feet high—at its greatest girth. The carcase was rather offensive and swollen with gases from decomposition, so that the notes taken by those who came near enough were rather hurriedly made. It is said that some of the Malacca officials had been to inspect it before, so that fuller details may be available.[31]

3.4 *The whale at the Raffles Museum in 1908*

3.5 *The whale at the Labuan Maritime Museum, Malaysia, in 2008*

The account of the whale's stranding and the subsequent action of the authorities, as published in the 1907 Report of the Museum and Library, is rather at odds with each other. In the 1907 Report, it was stated that a 'pagar' or line of stakes was built around the whale to prevent it from getting back to sea at high tide. The whale took a week to die, after which DPA Harvey, Resident Councillor Malacca 'caused the skeleton to be prepared to be conveyed to Singapore.'[32] This version gave the impression that the whale was actually still alive when Harvey and his team got to it, but that was not the case. The pagar was probably built to prevent it behind washed back into the sea while the labourers worked on the carcass.

The lack of space at the museum prevented the skeleton from being properly mounted for display and it remained in storage for the next 15 years. In 1907, the skeleton was finally mounted. Missing bones—a scapula, the 'hands', and several vertebrae and ribs—were modelled out of wood and plaster of Paris and the whole skeleton 'was suspended by steel ropes from the ceiling.' It was the ingenuous work of the museum's Chief Taxidermist, Valentine Knight —whom Hanitsch recruited during his 1902–3 furlough—and his staff. When unveiled, it was undoubtedly 'the most striking exhibit in the Zoological gallery.'[33] The museum now had on display, a specimen of the world's largest creature in its galleries. In May 1974, after the National Museum gave up its natural history collection to the Science Centre, the whale was taken down, dismantled into three pieces and sent by truck as a gift to the Muzium Negara (National Museum) in Kuala Lumpur. In exchange, the Malaysian would 'train Singapore museum technicians on the finer points of modelling and casting of exhibitions specimens.'[34]

expeditions that were to be a major feature of the museum's life in the decades to come. Alas, Davison's tenure was much too brief.

Haviland's Brief Interlude (1893–4)

Davison's sudden demise came as a shock to everyone and a replacement had to be found as soon as possible. Oldfield Thomas acted briefly as Secretary to the Committee before George Darby Haviland was appointed to take charge on 5 April 1893.[41] Thomas Quin took charge of the library. Haviland was a surgeon who joined the Sarawak Government as medical officer in 1891. Between 1891 and 1893, he was the second curator of the Sarawak Museum.[42]

He arrived in Singapore on 27 April 1893 and assumed his position during the library's 50th anniversary and the museum's 20th. Haviland stayed on for less than eight months but he was experienced in the running of museums and in his first and only Annual Report provided many observations and suggestions. Haviland thought that while the new building was much better than the old one, it had not been purpose-built or designed as a museum. The problem of lateritic dust, endemic in the old building at Raffles Institution—was still much in evidence here. Not only was it 'very injurious to the specimens' but it also gave a 'most unsightly appearance to everything which cannot be dusted frequently.'[43] The sparrows also created a nuisance as they found 'the inside of the Museum excellently adapted to their nests.'[44] And the termites, which had driven Davison out of his quarters, were 'still numerous in the walls of the building;' part of the roof had to be renewed as it was rendered unsafe by the termites.[45] And to top it all, the neglect of sanitation resulted in three cases of dysentery in the first part of 1893.

Haviland was particularly critical of the way in which the specimens were exhibited. He thought that the objects exhibited had to be 'carefully chosen and arranged so to teach their lesson readily.'[46] Cases suitable for museum display needed to be purchased as those in the museum were

> ...eminently unsuitable for purposes of exhibition and few are sufficiently well made to preserve good specimens, or even keep the red dust out; the greater number do not even keep out spiders. Owing to the absence of any suitable glass, there was no attempt this year to make better cases. Glass was ordered, and arrived in November, the greater part so broken as to be quite useless; it had, however, been insured.[47]

Haviland further added:

> If the Raffles Museum is to bear any comparison with other
> Museums in the Malayan region, it must have cases suitable for
> exhibiting and preserving the specimens exhibited. If shops, whose
> capital outlay is less than that of the Museum and whose customers
> are fewer than the visitors to the Museum, find it worth their while
> to buy proper cases for exhibiting their goods, it must be bad
> economy for the Museum to be carried on in the present way.[48]

Haviland's assessment of the museum's natural history collection is the
most detailed and accurate at this time and is thus worth reproducing
point by point:[49]

ZOOLOGICAL COLLECTIONS

Mammals

66. At any rate to natives, the mammals are the most interesting
and instructive portion of the Museum. Of the large mammals,
the Museum has as many specimens as its small size will permit;
as it is impossible to put them under glass, they deteriorate
rapidly. In the smaller mammals, the Museum is deficient; nor
is there a case fit to put them in. The monkeys form a most
interesting group on the Malayan region; of these there are a few
good specimens, but the cases in which they are do as much to
hinder them from being seen conveniently as to preserve them
from deterioration; and the same may be said of the carnivore. Of
squirrels, which are also numerously represented in this region,
there are but few specimens in the Museum.

67. The chief additions this year are 3 dolphins, 2 of them
purchased in Singapore, the other from Malacca; the skeleton of a
porpoise (*Orcella brevirostris*) purchased; the skeleton of a Malayan
bear (female) and a skeleton of a Malayan tapir (female), both
presented by the Director of Botanic Gardens. A few bats, new to
the collection, have also been added.

Birds

68. Several of the cases in which the birds are exhibited are
sufficiently well made to preserve the specimens, but they are not
suited either for purposes of exhibition or for the convenience of
those who wish to study birds. Horizontal rows of birds stiffly stuck
on varnished pedicels are instructive to few, and attractive to none.
Visitors prefer to see the showy and the common birds artistically

*3.6 Mounted
specimen of Fish
Owl, circa 1908*

mounted. Species which are attractive only to ornithologists, as from their rarity, or the peculiarity of their distribution, are more easily referred to, and more cheaply and efficiently preserved if kept in skin, and not put out for exhibition. Boxes have this year been made for keeping such skins. There are a good many duplicate skins of common species.

69. The additions to the specimens have been few; they include six species hitherto unrepresented in the Museum. The donors this year were the Hon'ble EE Isemonger, Mr HN Ridley, His Excellency the Acting Governor.

Reptiles and Amphibians

70. There are a number of these, especially of snakes in spirit, but many unfortunately have no record to the locality from which they came, and a large number are duplicates which should be disposed of. In this climate, spirit evaporates so rapidly that there is much waste in keeping more specimens than the Curator has time to properly attend to.

Fishes

71. A collection of stuffed fish made from 1882 to 1885 is in a good state of preservation. Spirit specimens of small fish, especially fresh water fish, might with advantage be added to the collection. There have been almost no additions this year.

Invertebrates

72. There are in the Museum a great many insects in store-boxes; these, however, are not well made, and the insects suffered very much from neglect at the commencement of the year, in some cases even the labels as well as the insects having been eaten up. There is at present no means of exhibiting any but a selection of the butterflies. Good insect cabinets from Europe are necessary if the insects are to be preserved and used either for exhibition or for reference. Insects are not easy to preserve in this damp climate, and it is cheaper to take proper steps for preserving those wanted than to be perpetually collecting, sorting and replacing.

In this extremely lengthy report, he made a very important point regarding mounted specimens for display, and non-mounted specimens for scientific study. While the public will be delighted and impressed by specimens mounted artistically, these may not be suitable for scientific study since specimens on display can hardly be studied closely by scientists:

> The standards of reference for scientific purposes should be made of use to those who wish to work at scientific matters. To exhibit all the specimens required for this purpose is to render them less accessible for reference than if carefully arranged in cabinets and boxes, to confuse the ordinary visitor, and to waste a quite impractical amount of space and money.[50]

This was the first emphatic statement by a Curator of a need to differentiate between a scientific reference collection and a display collection. Although it would take another 80 years for it to be recognised as such, the seeds of a Zoological Reference Collection had thus been sown.

The Committee was keen for Haviland to take on the post of Librarian and Curator on a permanent basis. However, following Davison's death the year before, the salary for Librarian and Curator had been reduced; Haviland would agree to stay only if the pay returned to its previous scale. The Committee unanimously recommended this move as they believed that they had in Haviland 'a Curator of exceptional value,' but this appeal was rebuffed by the Straits Government. Haviland resigned on 6 January 1894.[51] Thomas Quin, who took over as Acting Secretary and Curator, issued a press statement to inform the public that Haviland's resignation 'was not due to any difference of opinion' between Haviland and the Committee, but because Haviland was unwilling to accept the salary package he was offered.

Despite the transitory nature of the leadership at the museum, 62 specimens were added to the collection over the year, and the museum had 74,813 visitors. Alas, during Quin's management, the museum was broken into on 12 May 1894 and thieves made off with 'a portion of the replica of the Raja of Perak's Regalia.'[52] The Police failed to track down the perpetrators. Quin acted as Secretary and Curator for year and resigned in March 1895 and John Graham was appointed to replace him.

3.7 Hanitsch, with his signature pince-nez

The Great Hanitsch (1895–1919)

In their search the Committee now looked to Britain for a proper permanent replacement. Some years back, they had approached the British Museum to help identify possible candidates; when word got out that the Raffles Museum was looking for a new Curator, more than 60 persons applied. The Committee was looking for

> …a young man with plenty of energy and with good health and common sense. He should have some practical experience of the arrangement of a museum and the classification of the contents and an intelligent general knowledge of the various branches of science and natural history[53]

The successful candidate was Dr Karl Richard Hanitsch (1860–1940), a German-born zoologist with a doctorate from the University of Jena. In 1886, Hanitsch left Germany for Britain to improve his prospects. For a while, he taught in a number of English schools in

Bournemouth and Durham,[54] where he received board and lodging but was offered no remuneration. In fact, he was so poor he had to pawn his microscope to make ends meet. In 1887, things took a turn for the better: he received an inheritance when his aunt died, enabling him to redeem his microscope, and was appointed Demonstrator of Zoology at University College, Liverpool, with a salary of £80 per annum.[55] In 1892, he met and married Ethel Vernon. With his pinchnez, floppy head of hair and thick moustache, Hanitsch looked every bit the serious scientist. He arrived in Singapore to take up his position as Secretary and Curator on 1 July 1895.[56]

The First Year: Making Up For Lost Time

Hanitsch's youth and energy—as required in the job description—were swiftly displayed. Concerned that the museum had been without a permanent curator for 18 months, he prioritised 'overhauling, cleaning and arranging the large collection of stored zoological specimens.'[57] Realising that the museum's significance corresponded directly to its usefulness as an attraction and educational tool, he spent much of his time ensuring that exhibits on display were laid out scientifically, logically and aesthetically. As a scientist, Hanitsch grew concerned when he discovered that many of the specimens in the collections were not tagged with basic information, like where or when they were collected, thus rendering them 'almost useless for scientific purposes.'[58] He spent time uncovering the numerous collections that had been stored away and discovered a large collection of shells from the storeroom, which he proceeded to clean up and display, and commenced work on 115 boxes of insects languishing in the store.

His next task was to invest in new display cases with proper glass and brass fittings to better keep dust out and exhibits safe. Ten old cases were stripped of 'superfluous woodwork' and had new glass put in, making them 'more suitable for exhibition purposes.'[59] Hanitsch also ordered nine large cases from a Chinese carpenter, who constructed them based on a European pattern supplied by Hanitsch. All glass- and brass-work for these cabinets had to be ordered from Europe. He also ordered the construction of two aquaria and one vivarium. He would order more new display cases the next year, which allowed him to display even more material— sponges, *colenterates*, *echinoderms* and *crustacea*.

Amidst all this activity, Hanitsch also found time in his first year to organise several expeditions to Pulau Blakang Mati (Sentosa) and Pulau Brani to collect marine invertebrates at low tide. Several expeditions were also undertaken to Bukit Timah where spiders, insects, and reptiles were captured.[60]

After a year on the job, Hanitsch came to the same conclusion reached by his predecessors: too much of his time was being taken up in unproductive library-related work. His time should instead be spent on museum-related work, for example, investigating and studying the fauna in Singapore which he considered 'among the richest in the world.'[61] He suggested dispensing with the lending library and forming a proper reference library with the 13,000 or so books that were most suited. Doing so would free up the space for museum display, since the ground floor spaces were best suited for adaptation for museum use.

The problem of time management, however, was not so easily solved. Approximately 20 years later, in his 1913 Annual Report, he would continue to lament:

> In the review of the work done and left undone during the past year, one is invariably much tempted to bewail the variety of duties which have to be performed, the chief cause of complaint being that the Head of the Department has, besides the scientific work of the Museum, the control of an ever growing Lending Library, and also that the time available for the Museum is divided between two large collections in Zoology and Ethnology, and smaller ones in Botany, Geology and Numismatics.[62]

A Space Crunch...Again

At the end of 1896, Hanitsch commenced a complete rearrangement of the exhibits, moving the birds and reptiles into the Mammalia Room and putting the mammals in the Mineral Room, which was the 'best-lighted room of the building'. He also reorganised the Ethnological Collection, which had hitherto been scattered throughout the building and exhibited it in the former Bird and Reptile Room. Once the exhibits were located in what Hanitsch thought were the best locations for them, he then spent time overhauling the exhibits themselves. Systematically, Hanitsch dealt with each collection, replacing old specimens with better new ones, or inserting new specimens when they became available and added to the story the display told.

While rearranging all the exhibits, Hanitsch came to yet another conclusion his predecessors Davison and Haviland had arrived at: there was simply not enough space in the building for both a library and a museum. He had just ordered seven new cases for the Invertebrate Room, which he had to close and use as a workshop for the naming and mounting of specimens. It was clear to Hanitsch that even though the building was barely a decade old, he had run out of space. In December, he had eight additional large cases and a 'considerable number of stands and tables' ready for use. He closed the museum for three days and thoroughly reorganised the galleries as follows:

> The former Mammalia Room was given up to birds and reptiles, allowing them thus more space and light than they had formerly; some of the mammals were placed in the best-lighted room of the building, the former Mineral Room, and the remainder in the adjoining part of the gallery, assigning to the minerals a corner in the gallery where they find at present sufficient accommodation. Thus the former Bird and Reptile Room was cleared to receive the ethnological collection which previously had been somewhat scattered round the gallery. Although the room itself is, from the want of light and ventilation, the worst in the building, still it will, when re-furnished, be more suitable for its present purpose than the gallery.[63]

Space, or rather the lack of it, was a major preoccupation for Hanitsch, who was forced to shut down two exhibition rooms so that they could be used as a store and a laboratory:

> The enlargement of the Library and Museum building itself is also a matter for immediate consideration. The bookshelves are nearly full, there is little room for additional shelves, and in the Museum the rooms are crowded, the Mammalia Room especially so. The *mammalia* have apparently not had sufficient space for years, and in consequence, have suffered much from visitors, and even the present new arrangement is not quite satisfactory. No objects in the Museum should be exposed to the touch of visitors, they should either be under glass or so railed off from the public that to touch them would be impossible. But no space for such an arrangement is available in the present building.[64]

Of course, not everything should be displayed. If sufficient storage was made possible, reference specimens could be stored economically. It was, Hanitsch argued, 'neither desirable nor possible to have all specimens exhibited in the galleries' since the same material, 'well arranged in storerooms, would be of the utmost scientific value.'[65]

Within the first years of his tenure, Hanitsch developed a regime designed to keep the museum up-to-date, aesthetically pleasing to visitors, and scientifically significant to the ever-increasing numbers of visiting scientists and naturalists. While the work of the museum was all-consuming, Hanitsch was forced to spend a large chunk of his time running the library as well.

Money Matters

In 1899, the Straits Government approved the placing of the office of Librarian and Curator on the permanent establishment. This meant that Hanitsch's salary would henceforth be paid directly from the Colonial Treasury and not from the annual grant to the Raffles Library and Museum. The grant was accordingly reduced to $4,255 per year, down from the $9,000 it had previously been.[66] This was paltry by any account. Hanitsch compared the operating budgets of the Raffles Museum to those of the Perak Museum and the Federated Malay States Museum in Selangor and concluded that they were much better funded even though Singapore was the more important colony.[67] As Hellier drily noted in his Annual Report for 1901:

> As to the Museum, when it is remembered that Singapore is the centre of a region which ethnologically and zoologically is one of the most interesting in the world, it must be admitted that neither the present building nor its contents are worthy of the place.[68]

Funding for the library and museum had long been a sore point. While the Straits Government was initially prepared to provide an annual grant to the institution, it was not really prepared to finance it totally. The grant typically paid for the staff of the library and museum with a little left over for acquisitions of books and specimens. There was a modicum of income from subscribers to the library but this was meagre compared to the facilities and cost of maintenance. Indeed, in his 1902 Annual Report, Hanitsch worried about the lack of funds for the acquisition of specimens, especially ethnological ones and lauded the government for deciding against a brand new building for the museum but extending the present one, so that any surplus funds could be applied to acquisitions and new display cases:

> The funds available for the purchase of specimens have been quite insignificant for the last ten years or more, most of the meagre income of the double institution being absorbed by the Library, and to have a good ethnological collection, even if housed in a plain building, would seem the most urgent requirement of the Museum. Efforts to secure such specimens may be too late even in the near future. It is not too much to say that the Museum at present can only give an impression of poverty to the European visiting it, instead of conveying to him some idea of the vast resources of the Malayan region or of the wealth of the Colony.[69]

The government's annual grant, which started in 1887 with $10,000 per annum, was dropped to $9,000 per annum between 1890 and 1898. It was further reduced to $4,255 in 1899 when the salary of the Curator was paid directly by the government. It was raised by $500 to $4,755 in 1901, and then to $7,400 in 1902, and then to $10,000 in 1904. Nine years later in 1913, the grant would be $12,000 per annum.[70]

Patch and Repair, and an Extension

Beyond the fact that the original McCallum building was getting too small for the Raffles Library and Museum, it was also falling apart in places. In 1897, the Public Works Department had to overhaul the

3.8 *Exterior of the Raffles Museum in 1908*

3.9 *Hanitsch in the curator's office at the museum*

entire roof of the building as it had been 'leaking for some years.'[71] Work was also needed to improve the lighting and ventilation of the Bird and Ethnology Rooms. In 1898, the wooden floor of the building's west wing was removed to rid the place of termites once and for all, and the floor was filled up with soil and cemented over. The circular gallery was also improved when an iron spiral staircase was installed in place of the 'old rough wooden ladder', and the ceiling—another victim of the termites—was replaced. Work to incorporate the old Curator's quarters—abandoned by Davison after the termite infestation of September 1891—into the rest of the building began in June 1899. These alterations were completed in October that year and the additional space was used as a store and workrooms. Unfortunately, the repairs to the various ceilings proved to be only temporary solutions as the termites were still 'as active as ever.' The ceiling of the Reference Library on the ground floor was 'quite rotten', and that of the Circular Gallery was 'in parts hanging down in shreds.'[72]

After years of complaints about the lack of space, a proposal was put forward to erect a brand new building for the museum on the site intended for the Diamond Jubilee Memorial Hall (later occupied by the Victoria Memorial Hall). The estimated cost of this new building was $270,000 but this was considered excessive and was rejected by the Legislative Council. Nonetheless, the Council approved a vote for a sum of $50,000 to be applied for the extension of the present building.[73] In 1901, when Hanitsch went on long leave, the Acting Curator M. Hellier was informed that the plans for extension had been abandoned and the plans for a brand new building were back on the table. Hellier was pleased. Like many others, he felt that the two institutions should be separated and that the museum needed a much more capacious building than the library, which could continue to occupy the building on Stamford Road.[74] In the meantime, the ceiling of the Central Gallery was in need of repair again and the Gallery was closed for a month when these were done.[75]

After Hanitsch returned from long leave in 1902, he retained Hellier for a further three months to re-label a portion of the collection and to prepare 'descriptive popular labels for many of the better-known mammals.'[76] Hanitsch was also informed that the $50,000 vote had not received the approval of the Secretary of State for the Colonies. In any case, the Government Architect was asked to prepare new plans for the extension of the building. Six large rooms would be added to each end of the building (three on the ground floor and three more above). Hanitsch estimated that 'this additional space would be sufficient for about 20 years.' However, these proposed changes would cost $63,000.[77] In 1903, the Legislative Council once again revived the 'new building'

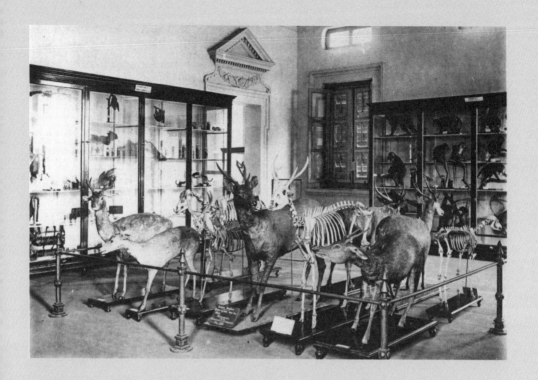

3.10 Two views of the zoological display at the Raffles Museum, circa 1910

proposal and approved the sum of $300,000 for the new museum building with $60,000 to be spent in 1904.

In 1904, Hanitsch reported that the government had increased the Museum Grant from $67,400 to $100,000, and that it had finally decided to extend the building, rather than construct a new one 'according to certain plans which [he] submitted five years ago'.[78] Work on the building commenced in October 1904 and was expected to be completed by the end of 1905. It was estimated that the extensions would cost the government $80,000.[79] The builders worked assiduously to get the building up and completed their work in the middle of 1905, ahead of schedule. Hanitsch, half-relieved and half-elated noted:

> The new building was finished towards the middle of the year and provided urgently needed accommodation for the rapidly growing collections, and spacious workrooms. The ground floor consists of seven large rooms, two of which were set apart for the Library, and the remaining five for the Museum as workrooms and store rooms. The upper floor contains a large central gallery measuring somewhat over 171' by 35' with a smaller room at either end, measuring over 40' by 30'.[80] The government voted a special sum of $8,900 to furnish the new rooms.

Reorganising and Overhauling the Exhibit

When the new extension was completed in 1906, Hantisch once again found himself reorganising the exhibits as he did in 1896. In particular, he moved the entire zoological collection to the new building in November 1906 but kept the ethnological collection in the old one. This new arrangement proved popular and when the new gallery was open to the public in time for Chinese New Year in 1907, 'both the new and the old galleries were filled to over-flowing by dense masses of Chinese holiday makers'.[81]

The work of maintaining and improving the displays was a never-ending one, given the problem of lateritic dust, rising damp, termites and the humid tropical weather. By 1910, for example, the showcases needed repainting, and it was reported that the taxidermists' 'time was occupied in keeping the present collections in order'.[82] This was because the peons were 'not skilled enough to clean specimens by themselves' so that the 'work had to be done by the Taxidermist himself or under his personal supervision'.[83] As Hanitsch noted in 1911, the collections, which filled '63 large wall cases, 102 table cases, and 48 smaller show cases' plus 'hundreds of store boxes' were now so extensive that 'a very considerable part of the time of the small staff is occupied in keeping them in order'.[84]

3.11 The circular lobby of the museum, circa 1910

Improving the Zoology Reference Collection

Hanitsch gave the highest priority to the zoological exhibits partly as a matter of professional interest, but mainly because he knew that it was this collection that attracted the greatest number of visitors. Most of the local population were still illiterate and were less attracted to the ethnological material, but there was something primeval and palpable about gazing at these magnificent specimens of nature at such close range that had people returning to the museum again and again. As Hanitsch observed, 'A Museum is of little use without its visitors, and the Raffles Museum can certainly not complain of the lack of interest, at least on the part of the natives.'[88] It was thus with great relief that he was able to move the entire zoological collection to the new extension of the museum and display them to maximum effect:

> The greater portion of the zoological exhibits was housed in twenty new cases, the Mammals and Birds receiving nine each, the Reptiles two. The cases are constructed of polished teak and plate glass and were supplied by Messrs John Little & Co. Sixteen of them are of uniform size, 14 feet by 9 feet by 2 feet 6 inches, with four doors each, the dimensions having been taken from standard cases in the British Museum, Natural History. For the man-like Apes a specially large case was constructed, measuring 22 feet by 10 feet by 2 feet 6 inches, with five doors. Most of the Fishes and Invertebrates were left in their old cases: only the Crustacea received two new wall cases, and the Shells and Insects several new table cases.[89]

Hanitsch managed a healthy relationship with many of the region's museums as well as those in India and in London. Through this network, he was able to trade surplus specimens or buy specimens from other museums to add to the collection. He was particularly concerned to 'complete' as much of each collection as possible by having a comprehensive set of specimens. Hanitsch was thus delighted when, in 1908, he managed to acquire 51 species of birds 'new to the collection' from the Selangor Museum partly through exchange and partly by purchase.[90]

The significance of the reference collection is alluded to in the Annual Reports of the Museum if one looks at the reports on the many scientific visitors Hanitsch lists each year. Much of Hanitsch's time was also spent showing these scientists, naturalists and scholars around and in sharing his knowledge of the collection with them. Indeed, they sometimes took up too much of his time. As he noted in his 1913 Annual Report:

> Much time is also taken up by scientific visitors, and however welcome their visits are in many instances, in other cases their desire to make a short stay in Singapore as profitable as possible, combined with much helplessness due to lack of local knowledge, is frequently a great tax on the time and resources of the Director and his assistants.[91]

Other Developments

In the meantime, a significant administrative development occurred in 1908 when the title of 'Curator and Librarian' was changed to that of 'Director'.[92] Hantisch was thus the first Director of the Raffles Library

Gifts from Johor

On 12 November 1909, Hanitsch received a message from the Sultan Ibrahim of Johore (through Walter Makepeace)[85] that he had shot an elephant near Senai in Johore and informing him that the Sultan wished to present it to the museum. This was no easy task, given the size of the elephant and the difficult conditions under which the taxidermists had to work. As Hanitsch recalled:

> ...after several days' strenuous work in the jungle on the part of the taxidermists the skeleton of the animal was secured. The elephant was a male, a magnificent specimen, though not quite adult, as the condition of the bones showed.[86]

The museum had long desired an elephant—either stuffed or skeletonized—and to have one shot and presented by the Johore Sultan made the specimen, Hanitsch thought, the 'most valuable' to be added to collection for the year. The Sultan had previously also presented a tiger and a black panther to the museum, thus continuing a family tradition of museum patronage going back to the Sultan's grandfather, Temenggong Daing Ibrahim. The elephant's skeleton was mounted and displayed by the end of the year.

In 1912, the museum received another huge gift, this time from the Johore government. This was the whole collection of the old Johore Museum that existed

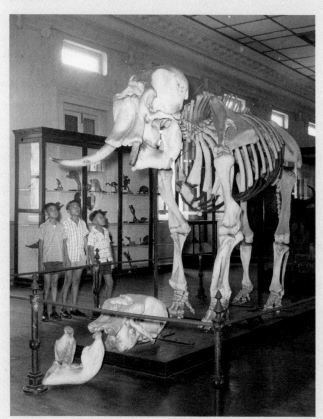

from 1904 to 1905. The collection consisted of '74 Mammal skins, 64 mounted Mammals, 480 Bird skins, 190 mounted Birds, 20 mounted Reptiles and about 200 Reptiles (snakes and lizards) in spirit.' Although most of the mammal skins were in poor condition, Hanitsch noted that the birds were 'in very good condition' and 1913 was dedicated to sorting and remounting this collection.[87]

3.12 Students from Tampines Primary School on excursion in front of the elephant skeleton

and Museum. One of Hanitsch's first endeavours as Director was to compile a Museum Guide. This guide to the museum's zoological collection was the first comprehensive and illustrated publication to be published. Hanitsch had spent most of his time 'outside the Library and Museum routine' working on this guide which was published in December 1908 as *Guide to the Zoological Collections of the Raffles Museum, Singapore*. It was 112 pages in length and contained 21 valuable plates. The photographs were shot by brothers Sim Boon Kwang and Sim Boon Eng who presented the plates to the Museum. It was priced at $1.00 and Hanitsch hoped that the affordable pricing would get 'the Singapore schoolboy' interested in the collections and purchase a copy.[93] Alas, Hanitsch was much too optimistic as sales proved less than brisk.

Maintenance and More Extensions

Building maintenance took up a considerable amount of time and seriously taxed the museum's annual grant. As we noted earlier, the building was constantly in need of repair, besieged as it was by the twin perennial problems of termites and the humid tropical weather. In 1910, general repairs were carried out throughout the buildings in an attempt to sufficiently weather-proof and burglar-proof the premises. The showcases constantly needed upkeep and repainting as well. In 1911, the entire building had to be 'repainted and colourwashed' due to the 'extensive ravages caused by white ants to the woodwork of doors, windows, floors and ceilings.' While this was taking place, there was 'a never-ending, moving about and cleaning up of show cases and specimens.'[94]

In 1913, an application was made to build a separate library building 'on the vacant plot of ground between the present building and the Bishop's house.'[95] The 1905 extension had been sufficient for less than a decade, and not for 20 years as Hanitsch had predicted. However, this plan was abandoned and it was decided that 'the accommodation required by the Library shall be provided by an extension of the new block…in the direction towards Tanglin, at a cost of about $75,000.'[96] With this new extension, the ground floor of the old building which was occupied by the library, would be freed up for the museum's collections.

Construction of the library extension commenced in August 1914 but work was slow as ground preparation was hampered by 'huge boulders of sandstone…imbedded in the clayey soil.'[97] The extension was finally completed in August 1916 and the entire library was moved over within a week. It was open to the public on 25 August 1916. Everyone agreed that the building, being 'bright and spacious' was 'much more comfortable and attractive' than the old library. The government provided $4,000 to furnish the new library and for showcases for the museum. Seven years after the first repainting in 1911, the whole building was once again colour-washed and repainted, inside and out, and the entire ceiling of the Animal Gallery renewed.[98]

The Passing of an Era

As Gretchen Liu noted, Hanitsch's energy began to flag towards the end of his tenure, quite possibly due to his constant battles with authority to get more funds for the museum, fighting the tropical weather in the upkeep of the buildings and collections, and at the same time, trying to continue furthering his own research. Indeed, he spent his long sabbatical leave on a prolonged busman's holiday. In 1915, when he went to Oxford for eight months, he spent his time 'engaged in working up the Malayan Cockroaches (over 200 species).'[99] For Hanitsch, the Raffles Library and Museum were his life's work, and as he approached retirement age, he worried about his future. He would, after all, turn 55—the age of retirement—on 22 December 1915. In March 1916, Hanitsch wrote to the government proposing his retirement. Apparently, a successor had been found but as the First World War was still ongoing in Europe, it was decided that a change of directors would only be effected after the cessation of hostilities.[100]

3.13 Front cover of the Guide to the Zoological Collections of the Raffles Museum, Singapore (1908), by Richard Hanitsch

One of Hanitsch's last major contributions to the museum was the establishment of his Singapore History collection. While the museum had a miscellany of artefacts pertaining to the history of Singapore, Hanitsch seized on the excitement and occasion of Singapore's

3.14 One of the 21 plates from Guide (1908): the scaly anteater

*(Above) 3.15
Postage stamp
commemorating
Hanitsch's visit to
Christmas Island
in 1904. The
stamp was issued
as part of the
'Famous Visitors'
series in 1978.*

*(Below) 3.16
1962 stamp
featuring the well-
known Harlequin
Rasbora
(Trigonostigma
heteromorpha),
first described
from Singapore in
1904*

centenary celebrations in 1919. He suggested the 'forming of a collection of portraits of distinguished residents of Singapore of the past and present, of plans and pictures of old Singapore, and of other subjects of historical interest.'[101] To this end, Hanitsch personally appealed to the press, companies, officials and individuals for cooperation, and while the response was not 'overwhelming, yet the beginning of a local Valhalla has been made.'[102] A number of portraits were promised, and Dr Gilbert E Brooke (1873–1936) 'handed over to the Museum a number of old plans of Singapore and interesting records which he had discovered in the Record Room of the Colonial Secretary's office.'[103] Brooke, who was then Chief Health Officer of Singapore, was in the process of researching, writing and compiling—together with Walter Makepeace and Roland St John Braddell—the monumental two-volume work, *One Hundred Years of Singapore.*[104]

Hanitsch left Singapore in April 1919, aged 59 and having served 24 years as the head of the Raffles Library and Museum. It was most unfortunate that a person of Hanitsch's capability, industry and loyalty was not more recognised in his lifetime. His German descent was, unfortunately, to blame for this. When he emigrated to England in 1886, anti-German sentiment in Britain was already high.[105] Things got worse after the war even though Hanitsch became naturalised as a British citizen in 1910.

Hanitsch retired to Oxford to be near to the famous Hope Entomological Collection at Oxford University. In 1919, he wrote to the government seeking help in finding employment either at home or abroad as he was concerned that his pension would be insufficient for his family's needs, especially since his two youngest children were only 8 and 6 years old.[106] The reply from the Colonial Office was terse:

> Sir, I am to acknowledge the receipt of your letter of 30th July, applying for further employment either in this country or abroad, and to inform you that his Lordship fears that he can hold out no hope of being able to offer you re-employment in the Colonial Service in view of your age. Lord Milner also regrets that he has no patronage in this country, and he cannot assist you to obtain employment here.[107]

At Oxford, he continued his research, spending a few hours every morning at the Department of Entomology right up till his death on 11 August 1940 at the age of 80. In March 1935, Oxford University, in recognition of Hanitsch's contribution to science, awarded him an honorary MA, a rare honour since he was not an Oxford graduate.[110] In 1978, Christmas Island issued four-stamp series featuring 'Famous Visitors'. The 50-cent stamp—the stamp of the highest denomination—commemorates Hanitsch who had visited the island for five weeks in September-October 1904 with Henry Nichols Ridley, Director of the Botanical Gardens.[111] Alongside a picture of Hanitsch, with his trademark pince-nez, white drill tunic and topee hat, is a picture of *Ducula whartoni*, the large Christmas Imperial Pigeon which had been described by Richard Bowdler Sharpe in 1887, the year Raffles Museum moved into its Stamford Road home.

Hanitsch's long tenure as Curator and then Director of the Raffles Museum allowed the institution to settle into its foundations, find its feet and grow. With his scholar's knowledge, scientific mind, and organisational talent, Hanitsch put Raffles Museum on the map as one of the most important institutions of its kind in Southeast Asia. Bashford Dean, Professor of Zoology at Columbia University toured several 'Asiatic Museums' in 1907 and wrote favourably of the Raffles Museum. Noting that it was 'too small for its needs', he singled out the collection of insects as being important and the magnificence of some of its specimens, especially the orang-utans, and the rarity of the dugong exhibit.[108] In closing, Dean attributes the success of the museum to the 'labours during the past dozen years of the Director, Dr R Hanitsch.'[109]

CHAPTER FOUR

From Strength to Strength
1921–42

KARL RICHARD HANITSCH'S 24-year tenure as Curator and then as the first Director of the Raffles Museum was unprecedented and would not be repeated. Over the next 22 years, the museum was to grow even more rapidly and become more important both in the public and the scientific world, and it was to be led by three men who were acknowledged giants in their respective fields: John Coney Moulton, Cecil Boden Kloss and Frederick Nutter Chasen.

4.1 John Coney Moulton (1886–1926), Director of the Raffles Museum and Library from 1919 to 1923

Laying Down The Law: John Coney Moulton

John Coney Moulton (1886–1926) was both a scientist and a military man. Born in St Leonards, Dorset, West Country, he was educated at Magdalen College, Oxford where he was Fellow Commoner from 1905 to 1908. In 1908, at the age of just 22, he left for Sarawak to become Curator of the Sarawak Museum, succeeding John Hewitt. When the First World War broke out in Europe in 1914, Moulton enlisted with the 4th Wiltshire Regiment and served first in India between 1915 and 1916, and then as a staff officer in Singapore between 1916 and 1919. In the course of his service, Moulton rose to the rank of Major.

When he was appointed Director of Raffles Museum and Library he resigned his commission. Moulton was interested in entomology, birds and mammals and in 1911, founded and edited the *Sarawak*

Museum Journal. Moulton's appointment took effect from 8 July 1919.[1] However, Valentine Knight, the Assistant Director, was to act in his place till December 1919 as Moulton had been granted eight months of unpaid leave following his discharge from the army. During this time, the government added three posts to the permanent establishment: Assistant Director, Librarian and Taxidermist. This meant that the salaries of these office-holders would be paid out of colonial government funds and not from the annual grant of the Raffles Museum and Library.[2]

The Same Old Problems

The approval of the appointment of a fully-qualified Librarian to take charge of the library freed the Director of the tedious routine work involved in running the lending library. With this, Moulton saw himself as Director of the Museum first and Director of the Library second. He thus reversed the order of reporting in his annual report, placing the museum before the library. From this point on, he was Director of the Raffles Museum and Library and the institution was accordingly referred to by this name henceforth. The committee also approved the appointment of an extra clerk and two Dayak collectors for the museum, thus enhancing its capability to build up its collections further. During Moulton's first year on the job, minor repairs were made to the floor and roof of the buildings and new windows were installed in the Ethnological Gallery and Clerk's office as these rooms were very dark.[3]

The person recruited for the post of Librarian was James Johnston from Scotland. He assumed duties on 10 May 1920. Knight, who had reached the pensionable age of 55, had his service extended on account of his valuable work, and the Secretary of State for the Colonies gave approval for the engagement of a new taxidermist. The person chosen for this post was Frederick Nutter Chasen, the Assistant of Castle Museum in Norwich. He arrived in early 1921. With his new team in place, Moulton could, as he put it, get on with the 'somewhat formidable task of grappling with arrears of reorganising and re-arranging both Institutions…vigorously and, perhaps for the first time in the history of this Department, under promising conditions.'[4]

Moulton highlighted the problem of space once again in his 1920 Annual Report. Although the library was presently occupying its own wing at the rear of the building, Moulton argued that more space was needed for the museum and that it was necessary for a new library building 'facing Fort Canning Road and immediately behind the present building.' Ground between the museum and St Andrew's School was also reserved for future extensions.[5] The current buildings were repaired, the wood-block pavement at its main entrance was replaced with a cement floor, and public lavatories were installed on the premises, while electric lights, fans, and telephones were installed in the offices of the Director and Librarian.[6]

The appointment of a Librarian on permanent establishment was, for Moulton, the first step in the eventual separation of the two institutions. The question of separating the library from the museum was once again discussed by the committee but it was decided that no further recommendations on this matter be made to the government till the library's new building had been completed.[7] In the meantime, Moulton conducted a thorough review of the collection and the museum's collection policy. He was, after all, the first director of the museum who could afford the luxury of devoting all his time to the museum's work and not worry about the library's operations.

Moulton the Organiser

It will be recalled that back in 1889, it was William Ruxton Davison who urged that the museum limit itself in collecting specimens from the Malay Peninsula and adjacent lands within Alfred Wallace's 50-fathom line. Davison's injunction was, apparently, not heeded by his successors; Moulton found himself revisiting this issue in 1920 and once again laying down a geographical limit to where specimens ought to be collected from:

> With the limited space available and realising the futility of attempting to exhibit more than a ridiculously small collection to represent the fauna and ethnology of such widely separated countries, it was decided to limit the collections of the Museum as far as possible to the true Malaysian area. For this purpose I have defined this Malaysian sub-region as a rough square bounded by Lat. 10°N and 10°S, Long. 95°E and 120°E. This area includes the Malay Peninsula, Borneo, Sumatra, Java and adjacent small islands, but excludes Palawan, Celebes, Lombok and islands to the East of Wallace's line. Collections from, say, the Malay Peninsula lose half their instructional value without comparative collections from the other Malay countries, which, while so closely similar in many respects, differ in an interesting manner in so many other features both zoological and ethnological. It is, of course true that non-Malaysian elements are to be found in parts of this Malaysian area while true Malaysian forms have entered their range beyond the above-mentioned limits of Malaysia. These boundaries are therefore to some extent arbitrary.[8]

Moulton observed that 'Asiatic visitors take far more interest in the products of their own country, which are familiar to them, than they do in strange things from other countries unknown to them' while the European visitors 'take a natural interests in the strange products of their temporary home.' Bearing this in mind, it was 'thought advisable to aim primarily at the formation of a general Malaysian Museum' and collect material from further afield only 'if and when space is available.'[9]

The Reference Collection

Moulton's most important edict was related to the reference collection and the proper classification of specimens. Any museum that 'makes any pretence of being a scientific institution must' Moulton argued, 'adopt as a basic principle that science is "ordered knowledge" and as such, prioritise the "work of classification."'[10] He went on:

> For the natural history side of the Museum, scientists at work since the days of Linneus, have gradually evolved a very comprehensive system of classification, wherein every living creature finds a definite allotted place in a natural scheme, indicating so far as our present knowledge permits true relationship between all living things. Every natural history specimen in a well-ordered Museum must be fitted into its correct place and given its correct name. Now just what that place or name may be is unfortunately too often a matter for debate, so that a name or place for any particular species

4.2 *The Reference Collection Room of the Raffles Museum in 1931*

accepted as correct, say 20 years ago, very likely gives way to some other name, as the result of some recent worker's researches. To keep up to date in this respect is a big task. It can only be done with the assistance of specialists from all over the world. One of the first endeavours therefore of the director has been to work up a connection with leading Museums and specialists who are likely to be of use in this way. So far the responses to our requests for assistance have been eminently satisfactory. Collections have been sent to specialists in England, Finland, France, India and America for identification, and negotiations are proceeding with specialists in other countries. A considerable correspondence has also taken place with Institutions nearer at hand, viz., Museums and Societies in the Federated Malay States, Java, Sarawak, Siam and the Philippine Islands—all with the object of developing the classification work of the Museum collections.[11]

In his effort to systematically rearrange and reclassify the specimens, Moulton laid out a plan to first rid the galleries of anomalies so that sharks, crocodiles, whales and birds were not displayed together. The second step would be to re-catalogue all the specimens. While previous lists of mammals, birds and reptiles have been published, these 'are useless as catalogues' as 'no numbers have been given' to them in the past.[12] It was thus impossible to know exactly how many specimens there were in the museum. Moulton proposed a 'careful overhaul of each section' and the creation of detailed catalogues. Specimens that were old and had no data were to be eliminated, and the correct nomenclature would be used in a thorough relabelling exercise. By the end of 1920, Moulton reported that 'some progress on these lines had been made with the Mammals, Birds and portions of the Insect collections' and that the Ethnological Collections would be worked on in 1921.[13]

It was clear to Moulton that the museum's material should be divided into three series of collections. The first of these would be the Exhibition Collection, consisting mainly of mounted specimens displayed for popular instruction; the second would be the Reference Collection, which should be on as large a scale as possible, but carefully classified, named and stored away for consultation by scientists or advanced students of natural history. The third series would be the Duplicate Collection, which should be stored away for use as replacements for damaged specimens or for exchange with other museums. While the Exhibition Collections would be of primary importance in Singapore, Moulton was anxious that surplus display specimens be eliminated so that the limited space might be better utilised. He allocated one

4.3 *Two views of the zoological displays
at the Raffles Museum in 1931*

room for the Reference Collections which had previously languished 'uncared for in the workshop…[a] prey to dust and neglect.'[14]

Beyond being a mere repository for unique specimens, Moulton felt that the museum should also be a centre for research. Moulton wanted to make the museum 'a live institution, attempting to take its place among the scientific outposts of the world in the general attempt to advance human knowledge.'[15] This required making the Reference Collection known and available to those who would study it. Moulton put it thus:

> It is obvious that any active Museum, which has the good fortune to be situated in an intensely interesting tropical region such as this, cannot rest content with a sedentary existence as a mere store-house or receptacle for things already studied and described.
>
> To prosecute researches and to assist the studies of others in any biological and ethnological matter connected with Malaysia is held to be one of the primary duties of this Museum, so far as limited staff and funds will permit. Native collectors have been engaged to procure fresh material from suitable localities in the neighbourhood of Singapore. The collections of the Museum are placed freely at the disposal of recognised scientists who ask for the loan of our material to assist them in their researches.[16]

Moulton proposed providing a laboratory for advanced students in the future, if space allowed. In the meantime, a useful Reference Library had been set apart from the general library's collection for use by the museum's staff and 'scientific visitors.' The final step would be for the museum to have its own official publication. Although the museum's staff had written papers, they were usually published in the *Journal of the Straits Branch, Royal Asiatic Society*.[17]

In his audit of the collections, Moulton concluded that there was 'practically no reference collections worthy of the name' among the zoological specimens. What was even more pressing was the need to re-label all the specimens according to the latest zoological nomenclature:

> The great activity now going on all over the world in revising zoological nomenclature, chiefly due to the introduction of trinomials, has had the further effect of putting out of date probably at least 90 per cent of the names given in these Museum lists. It will be necessary to re-examine critically every individual specimen in the Museum, re-name it according to modern ideas, re-label and catalogue it.[18]

To this end, Moulton had enlisted the assistance of Herbert Christopher Robinson and Cecil Boden Kloss, respectively the Director and Assistant Director of the Federated Malay States Museums in Kuala Lumpur, to work on the mammals and birds, subjects in which they are acknowledged authorities.[19] Moulton conducted what was up to this time, the most thorough audit of the museum's collections and put in place a systematic plan to bring everything up to date and on par with the best museums in Malaya.

In October 1921, the government approved new designations for Knight and Chasen who were now respectively Curator and Assistant Curator. Together with Moulton, they spent most of 1922 weeding out surplus and duplicate specimens and artefacts. A selection of non-Malayan weapons and pottery that was considered of insufficient interest was auctioned off for $165, as were some duplicate brassware and non-Malayan birds for $111. With the help of a paid assistant, Mrs D. Horton, the museum was able to make 'great progress... in the work of overhauling, rearranging, labelling and cataloguing the bird collection.'[20] It also enlisted the assistance of its Chairman, Sir John Alexander Strachey Bucknill, to prepare 'a small handbook of the common birds of Singapore.'[21] Bucknill, who served as Chief Justice of the Straits Settlements from 1914 to 1920, was a very keen ornithologist and lepidopterist and had, in 1900, already published *The Birds of Surrey*.[22] The Singapore handbook was eventually published in 1927, a year after Bucknill died in India.[23]

Robinson, who had been so helpful in working on the mammals and birds in 1920, once again assisted in identifying the whole collection of *Tupaias* (tree shrews), and even in renaming many other mammals, especially the rodents. Robinson also presented the Museum with '8 Tupaia, 78 Squirrel and 65 Rat skins' to help fill the gaps in the museum's series of Malaysian species.[24] Moulton was also pleased to report that at the end of 1920, the museum's bird collection comprised 'approximately 4,500 specimens' (about 1,000 mounted; 3,500 skins). During 1921, another 950 specimens were added, representing about 223 species and that excluding Passeriformes (perching birds), 'the total number of species now recorded from Malaysia is 458, of which 284 are represented in the Museum.'[25]

Moulton and Chasen were both preoccupied with organising and preparing exhibits for the 1922 Malaya-Borneo Exhibition in Singapore, which was declared open by the Prince of Wales. Knight then retired in May 1922. By the end of 1922, the general rearrangement of the museum, which Moulton had mapped out in 1919, was complete. Moulton also reported 'good progress with the reference collections.'[26] The bird catalogue had been completed 'to the end of the non-Passerine

4.4 *Malayan Long-Tailed Parroquet* (Palæornis longicauda), *drawn by C. A. Levett Yeats. Coloured plate from John Bucknill's* Birds of Singapore Island *(1927).*

4.5 *Malayan White-Collared Kingfisher* (Halcyon chloris humii)*, drawn by C. A. Levett Yeats. Coloured plate from John Bucknill's* Birds of Singapore Island *(1927).*

birds,' and new cabinets with 'proper glass-topped boxes' had been 'purchased for the bird-skin collection' which was now in good order.[27] In addition:

> The spirit collection has been thoroughly overhauled, rearranged and made available for study.
>
> Useful progress has been made in arranging, extending and naming the insect collections. The assistance of many specialists all over the world has been enlisted in this work. Papers have been published or are in preparation describing new species contained in these collections. As a result the Museum is becoming more widely known as a live institution, and an increasing number of inquiries have been received from many distant parts of the world concerning the fauna of Malaya.
>
> …The room set apart for the Reference collections has been improved by the addition of eight windows. Its proximity to the biological reference library housed in the next room makes it a useful room to study; it has been utilized by several people, working out their own collections or studying those in the Museum.[28]

Moulton was the right man for the job but he did not stay long enough. One of the highlights of Moulton's short tenure as Director of the Museum was his contribution to the founding of the Singapore Natural History Society in 1921. Moulton was its first President and Chasen served as Honorary Secretary. This society may be regarded as the predecessor of the Nature Society (Singapore) and the Malayan Nature Society. In 1923, he resigned to take up the post of Chief Secretary of Sarawak, a much more important position politically. Moulton did not live long after leaving Singapore. He died on 6 June 1926, at the age of 40, a few days after being operated on for appendicitis.[29]

Other Neighbouring Museums and Personalities

While Singapore was clearly the most important British colony in the region and the centre of trade and commerce, it still did not have a museum commensurate with its standing. Three smaller museums—in Taiping, Perak; Kuala Lumpur, Selangor; and Kuching, Sarawak—were better funded and were able to attract some very talented and capable naturalists as curators and directors. At this point, it would be useful to look at museum developments in other parts of Malaya and Borneo.

It is important to understand Moulton's anxiety in bringing the Raffles Museum up to date within the context of developments in the Malay States and in Sarawak. In any case, a quick look at the histories of these institutions show how interconnected they are and how they all have a direct connection with the Raffles Museum.

4.6 The old Sarawak Museum building, built in 1891

Sarawak Museum

Of the three museums under consideration, the oldest—technically speaking—is the Sarawak Museum, established by Rajah Charles Brooke in 1860, largely through the influence of the great naturalist Alfred Russel Wallace. However, it was not till October 1886 that a temporary museum was established at the Market Place along Gambier Street. A proper museum—the present one—was built and opened on 4 August 1891, and a publication—the *Sarawak Museum Journal*—produced from 1911.

The first curator of the Sarawak Museum had been E. A. Lewis who served from 1888 to 1902. Two of the curators of the Raffles Museum—George Darby Haviland and John Coney Moulton—had served there in 1891-3 and 1908-15 respectively, before their tenures in Singapore. Today, the museum is known as the Sarawak State Museum, and is an excellent omnibus museum with collections in archaeology, arts, ethnography and natural history. The natural history collection is housed in a building that used to be the museum's administrative wing.

Perak Museum

The Perak Museum in Taiping lays claim to being the oldest museum in Malaysia, having been founded in 1883 by Sir Hugh Low, British Resident of Perak. Unlike the Sarawak Museum, which did not have its own home for more than 30 years, construction on the Perak Museum commenced in 1883 with funds raised through the efforts of Low and Sir Frank Swettenham, Low's successor as British Resident of Perak and later Resident-General of the Federated Malay States.

It was only completed in 1886 due to a lack of funds. Its first Curator was Leonard Wray Jr. (1852–1942). Wray, the son of a British planter, was born in Perak and educated privately. In 1881, he joined the Public Works Department and was Superintendent of Government Hill Garden at Larut between 1882 and 1903. In January 1883, he was appointed Curator of the Perak State Museum, where he collected and prepared Perak exhibits for the Colonial and Indian Exhibition of 1886. In 1904, he was appointed Director of Museums for the Federated Malay States.[30] Wray published *Perak Museum Notes*, the official journal of the museum from 1893 to 1898. This museum survived the Second World War intact and is today known as the Perak State Museum.

Selangor Museum

The third museum of note is the Selangor Museum in Kuala Lumpur. The museum started informally in 1887 at the house of John Klyne,

4.7 The old Selangor Museum. Built 1898, its right wing was completely destroyed when it was bombed by an American B-29 bomber in the closing months of World War II. It was torn down in 1959 to make way for the new Muzium Negara.

Chief Surveyor of Selangor. Klyne had let part of his house be used as a museum till June 1888, when the specimens were removed to Batu Road, and then to the new Istana building near the Roman Catholic Church,[31] where it remained till 1906. It then moved to its own purpose-built premises near the Lake Gardens (where the current Muzium Negara stands).[32] In 1903, Robinson was appointed Curator. In 1908, he became Director of the Federated Malay States (FMS) Museums. He was later joined at the museum by curators Cecil Boden Kloss and Ivor Hugh Norman Evans in 1908 and 1917 respectively. When Moulton resigned in 1923, Kloss moved to Singapore as Director of the Raffles Museum.

During the war, one wing of the Selangor Museum was destroyed by an American B29 bomber on 10 March 1945 and its collection had to be moved to the Perak Museum. It continued to house historical and cultural relics till 1959, when it was torn down to make way for the Muzium Negara or National Museum. It was officially opened in August 1963.

Joura F.M.S. Mus.—Vol X   Pl. II

1. RANA PULLUS.   2. NECTOPHRYNE PICTURATA. X2.

4.8 Plate from Malcolm Arthur Smith's 1921 paper on new amphibians from the Malay Peninsula and Borneo. H. C. Robinson had arranged for Smith to work on the specimens.

Federated Malay States Museums

In 1904, the Perak and Selangor Museums were administratively merged to form the Federated Malay States (FMS) Museums Department under which the *Journal of the Federated Malay States Museums* was published. In curatorial terms, the two museums remained separate and their respective annual reports were published in the *Journal*. Wray was, on 17 March 1904, appointed the first Director of the FMS Museums and held this post till his retirement in 1908. His post as Curator of the Perak Museum was taken over by Frederick W Knocker. Wray was succeeded by Robinson, hitherto Curator of the Selangor Museum in 1908, and when Robinson retired in 1926, he was succeeded by Kloss, who had the unique distinction of being Director of the FMS Museums and the Raffles Museum and Library.[33]

Moving Collections Between Museums[34]

To a large extent, the research, study and writing of the natural history of the Malay Peninsula, Singapore and Borneo was driven by three men who ran these museums: Robinson, Kloss and Chasen. Indeed, it was Robinson and Kloss who brought the large reference collection of mammals and birds from the Selangor Museum to Singapore in exchange for the Raffles Museum's entomology collection.

Moving specimens between museums was not uncommon. Starting from the 1880s, when Wray was Curator of the Perak Museum, specimens—especially holotypes—were sent from the museums in Singapore, Perak and Selangor to the British Museum (Natural History) in London, the central depository for the British Empire's natural history collections. Colonial curators and museum directors sent, as a matter of course, all holotypes to London. At the same time, some of the best specimens were also despatched to London. Alas, any record of these transfers on Singapore's side are now lost to history. It is, however, possible to reconstruct the scale and nature of these transfers with reference to specific collections. Dr David R. Wells, who spent a lifetime studying bird specimens in the collections of the Raffles Museum and of the British Museum (Natural History) in London, was able to reconstruct how these transfers took place with respect to passerine birds from Malaya and Singapore:

> Basically, the…accession registers show there was small-scale movement of bird specimens from the Peninsular museums to London, intermittently, from the 1870s (starting with Leonard Wray as curator of the Perak museum sending most of his Perak mountains collections to British Museum (Natural History) ornithologist RB Sharpe). Some specimens also went…to the Rothschild private museum at Tring, UK. Transfer movements really got going when bird collecting really got going, i.e., with the arrival of Robinson.…the latter's first Malay Peninsula collection was made on an expedition from the Liverpool museum, before he actually settled in Malaya. Most or all of that collection ended up at the BM(NH) in 1905.
>
> The first really big (hundreds of skins) transfer to British Museum (Natural History) made after Robinson joined the Selangor museum staff in 1903 was accessioned in London in 1906. This included a lot of his (June–July 1905) Gunung Tahan expedition materials plus more from various sites on the Main Range, etc. A second biggish transfer (100-plus skins from various localities, i.e.,

not biased towards any one expedition) was accessioned in London in 1908. A third big one (hundreds of skins) reached London in 1910—again including specimens from various places but mainly a big share of the results of a 1909–1910 Federated Malay States Museum expedition to peninsular Thailand.[35]

Wells speculates that the large-scale transfers from the Selangor Museum may well have been necessitated by a storage problem, especially since Robinson, Kloss and Chasen had been collecting furiously and the museum would have been stuffed full of specimens. The interesting thing, Wells says, is that the 'big-scale transfers' stopped at some point, which suggests that the storage problem at the Selangor Museum was solved. Indeed, large-scale transfers of specimens did not begin again until after the birds and mammal specimens were moved to Singapore in 1927:

The scientific collection that we have here [in Singapore] hails from Selangor. It got moved here. Selangor Museum was started as a kind of an amateur thing. Then they started collecting birds around late 1890s and really got going when HC Robinson took over around 1902 or 1905. He and CB Kloss were the main drivers. They established proper scientific collections of birds and mammals. It was all really at the Selangor Museum in Kuala Lumpur until Robinson retired in 1926. At that point, the whole Museum system was revised and all the vertebrates came to the Raffles Museum in Singapore and Kloss became the Director, and all the invertebrates went north, to Kuala Lumpur. That's how this museum acquired this collection. A big collection arrived in 1926 [sic]. The other half is in Tring. The reason for that is that there was an arrangement with the Imperial Government and the Colonial Government that the FMS Museum Service, would, approximately every 5 years, would split what they had collected in that period. One half of it would remain in the country and the other half would be sent to London. So if you want the whole collection, you need to go to the two collections. That remains the case today. And that went on until after the Collection moved to Singapore. I think FN Chasen was the last Director to implement that split. One lot went on till about 1934, 35 or 36, then he would have been preparing the next consignment and that was overtaken. What happened to that material, I wouldn't know. So that's how the collection got effectively split in two halves. Selangor Museum ceased to have a vertebrate collection and specialised in invertebrates—insects and so on.[36]

Publishing Scientific and Reference Works

Herbert Christopher Robinson—the only director not to have worked full-time at the Raffles Museum in Singapore—had long been working on the first volume of his *Birds of the Malay Peninsula* (on common birds) which he published in 1927, a year after his retirement. In 1928, he published a second volume of this massive work, focussing on montane species. Robinson died in May 1929, unable to complete the project. F. N Chasen, then still Curator at the Raffles Museum took up Robinson's project and working on Robinson's notes, published a third volume (on sporting birds, birds of the shore and estuaries) in 1936. The fourth volume (on birds of the low country, jungle and scrub) was published by Chasen in 1939. Chasen was killed in 1941 and most of his notes and records, including his plans for Volume 5 were lost with him, although the coloured plates which Chasen ordered from artist H Grönvold did survive and were later used by Lord Medway and David Wells to complete the series in 1976.

4.9 *Herbert Christopher (HC) Robinson (1874–1929), Director of the Federated Malay States Museums from 1908 to 1926*

Robinson also had plans to work on a book of the mammals of the Malay Peninsula and upon his retirement, 'brought to England a large collection of Malaysian mammals.'[40] As we noted above, Robinson died within three years of his retirement and at the time of his death, he had done little work on this proposed work. Kloss, who had been his closest collaborator, had initially agreed to continue with the work.

It seems that Kloss had every intention of working on the mammals book because in 1927, he arranged to have the entire bird and mammal collection of the Selangor Museum transferred to Singapore 'in exchange' for the Raffles Museum's insect collection. He was able to do this as he was Director of both the Raffles Museum and the FMS Museums. At first blush, such a wholesale transfer of collections seems highly irregular and defies logic, but the logic is revealed in a brief statement on the transfer in which Kloss said:

> A feature of the year has been the transfer, with the sanction of the Government concerned, of the large study collections of mammals and birds belonging to the Federated Malay States Museums Department from Kuala Lumpur to Singapore: reciprocally, the study collections of insects formerly housed in the Raffles Museum are now accommodated in the Selangor Museum.
>
> By this arrangement these collections are now under the immediate supervision of the officer most qualified to deal with them.[41]

In the last sentence quoted above, Kloss was most probably referring to himself since he was indisputably, alongside Robinson and Chasen, one of the leading experts on the birds and mammals of the region. At the same time, Henry Maurice (HM) Pendlebury, the newly-appointed Curator of the Selangor Museum was an entomologist. With this transfer, the birds and mammals collection of the Raffles Museum grew tremendously. However, despite his intentions to publish the book of mammals, Kloss 'for various reasons was unable to do so.'[42] This collection was later passed to the British Museum (Natural History).

Cecil Boden Kloss:
The Systematist

Cecil Boden Kloss (1877–1949) was an enigma.[37] Little is known of his early life save that he was born on 28 March 1877 and came to prominence when he accompanied the American naturalist, William Louis Abbott to explore the Andaman and Nicobar Islands in December 1900. Kloss and Abbot sailed from Singapore on board the schooner *Terrapin* to the Mergui, Andaman and Nicobar Islands on a collecting and exploration expedition. Kloss published an account of this journey in 1903,[38] the same year he joined the Botanic Gardens service. In 1904, he became Curator of the Johore Museum but resigned in September 1905.[39] Kloss, whose name is of Dutch origin, joined the Selangor Museum in 1908 as Assistant to H. C.

4.10 Cecil Boden Kloss (1877–1949), Director of the Federated Malay States Museums and the Raffles Museum and Library from 1926 to 1932

Robinson and the two formed a formidable scientific and exploration team. Kloss remained Robinson's deputy till 1923 when he replaced John C. Moulton as Director of the Raffles Museum.

When Kloss took office, Knight had just retired and Chasen was appointed Curator, making him Kloss's right hand man. In all, Kloss served eight years as Director. He was also, quite possibly, the first white man to climb the three highest peaks in Malaya (Gunung Tahan), Borneo (Mount Kinabalu) and Sumatra (Mount Kerinci). Beyond the usual task of estate maintenance and alterations to the museum's buildings, Kloss and his team spent most of their time either collecting new specimens or systematically labelling and relabelling them according to the latest scientific standards. For example, Chasen reported that the whole of 1923 was 'devoted to a whole-hearted attempt to get the zoological collections, especially the birds and insects, determined in accordance with the latest researches of systematists.'[43] Chasen also noted the steady build up of the reference collections that were being housed and arranged in manner that made it easily accessible to scientific workers. It was also gratifying for the museum to know that it was more frequently being consulted by local residents 'seeking information from a commercial or scientific point of view.'[44]

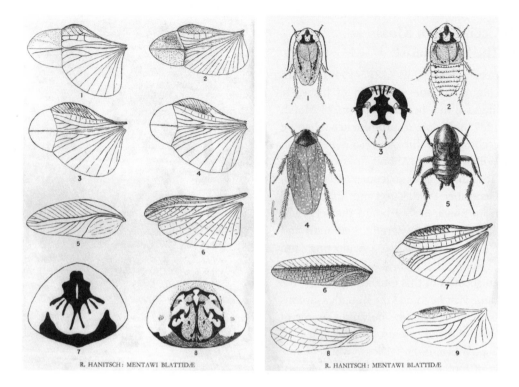

R. HANITSCH: MENTAWI BLATTIDÆ   R. HANITSCH: MENTAWI BLATTIDÆ

4.11 Two plates from Hanitsch's paper on cockroaches in the first volume of the Bulletin of the Raffles Museum *(1928). The wing of* Blattella ridleyi *(named after H. N. Ridley) was described from Singapore* (right plate, Fig. 6).

Kloss and his team developed a routine that helped keep the museum in the best shape possible. A perusal of the *Annual Reports* during Kloss' tenure will show that he and his colleagues spent their time between building maintenance and the periodic renovations, collecting expeditions, and organising and cataloguing the exhibits and specimens in the Reference Collection. In between, they would engage in scientific investigation of the specimens and publish papers describing their findings. It is thus significant that it was under Kloss' directorship that the *Bulletin of the Raffles Museum* was started in September 1928. The first issue featured, most appropriately, an article by Richard Hanitsch, the museum's first Director, entitled 'Spolia Mentawienasia: Blattidae.'

Kloss officially retired from the Raffles Museum in 1932 but left in November 1931 on long leave prior to retirement. He left as quietly as he came and retired in Lymington, England, where he indulged in his main hobby, yachting, and 'ceased for ever the scientific studies to which so far he had devoted his life.'[45] Edward Banks made this assessment of Kloss in the obituary:

> When he retired from Singapore in 1932 as Director of Museums, FMS and SS, Kloss was the foremost systematist on the Mammals

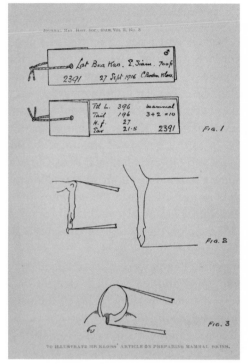

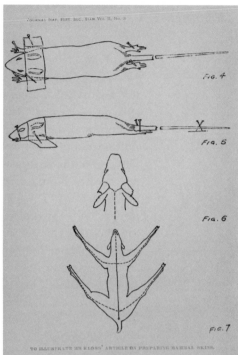

and Birds of South East Asia, with a long series of papers to his credit describing a host of new species and subspecies. Similarly he was an ardent botanist and his name is honoured in the genus *Klossia* (*Rubiacae*) and the many plants named after him specifically. He neither sought nor received scientific honours and I do not think he ever descended to popular science on a single occasion, though his work forms the purely scientific basis on which rests our knowledge of Malaysian Zoogeography...

Perhaps the key-note of Kloss was thoroughness, all that he did was done well, an accomplished traveller, ardent collector, skilful taxidermist, fine shot and a first class photographer, he carried out his affairs with zest and with confidence, aided by a wide knowledge and intellectual powers of a high order.[46]

4.12 Figures that accompanied Kloss's paper entitled 'Instructions for Preparing Mammal Skins' (1917). It was written for the 'use of those who wish to make collections'.

Kloss was the first and only person to hold the combined post of Director, FMS Museums and Raffles Museum and Library. After his retirement, this post was abolished and each of the museums reverted to having their own Directors. It was only on 1 January 1939 that the post of Director of FMS Museums was 're-created' with Henry Maurice Pendlebury (1893–1945), an entomologist and former military man, as Director.[47]

Frederick Nutter Chasen (1896–1942)

When Kloss retired, he was succeeded by Frederick Nutter Chasen. Chasen was born in 1896 in Norfolk and at age 16 articled as a pupil with Frank Leney of the Norwich Museum. After serving with the Norfolk Yeomanry during the First World War, he returned in 1919 to the Norwich Museum, where he cultivated his love of ornithology and museum work.[48] Two years later, Chasen was appointed Assistant Curator and then became Curator of the Raffles Museum after the retirement of Knight in 1922. In 1931, Chasen was appointed Acting Director of the Raffles Museum and Library, and his post as Curator devolved to Norman Smedley. Chasen was confirmed as Director on 28 March 1932 when Kloss retired. That year, the museum hired Michael W. F. Tweedie as Assistant Curator. He would be promoted to Officiating Curator in 1933 when Smedley retired on grounds of ill health, and later confirmed as Curator in June 1934.

Some time in 1930, Chasen had begun working on *A Handlist of Malaysian Birds*. This massive compilation of detailed scientific data on the birds of Malaysia was published in 1935.[49] Chasen followed this up with his monumental *A Handlist of Malaysian Mammals*, which he published in 1940.[50] In addition to these two important books, Chasen wrote profusely and published prolifically, averaging four papers a year.[51]

By this time, the museum's Reference Collection had become one of its most important assets. In 1934, the collection received a major boost with the addition of the H. C. Abraham collection of spiders which 'consists of many thousand of carefully labelled and recorded specimens' together with 'twelve volumes of field and colour notes and the elements of a formal catalogue.'[52] In 1935, the museum received a 'large comparative collection of marine forms' from Borneo from Henry George Keith of Sandakan and Edward Banks of Kuching.[53] The museum also purchased an 'extensive collection of birds and mammals from north-east Sumatra.'[54] In 1937, the American ichthyologist Albert William Christian Theodore Herre (1868–1962) visited Singapore and helped the museum identify 'the bulk of the undetermined material in the spirit room.' In the process, a number of species, new to the collection, were added.[55] The last major pre-War addition to the Reference Collection was a group of frigate birds and crustacea from the Christmas Islands donated by Carl Alexander Gibson-Hill of the Christmas Island Phosphate Co. in 1939.[56] Gibson-Hill would later join the museum as Curator and served as the museum's last expatriate Director.

The Beginnings of War

The largest transfer of all time to the British Museum (Natural History) took place in the 1930s when Chasen was director of the museum:

> By far the biggest transfer of all time, probably 1000-plus skins, spread right across the collection, i.e., not focused on any particular expedition, and with field dates going back to as early as 1907, was accessioned in London in 1936. This date fits with Chasen having completed and published his magnum opus, *A Handlist of the Birds of Malaysia*[57] and the guess is that, having finished his research, he was again relieving a storage problem. He would then have been working on the parallel project, his *A Handlist of Mammals*[58]...

We do not know if Chasen was planning a massive transfer of the mammal specimens upon the completion of his mammals book since war had already broken out in Europe, and the specimens were probably much safer in Singapore (at least in 1940) than in London.

As Britain entered the war in Europe, research activities at the museum slowed down although the museum staff continued improving the exhibition displays. However, Chasen worried incessantly about the type specimens in the Raffles Museum's collection. He had been home on furlough from February to December 1939, and would have been in England when Britain declared war on Germany on 3 September 1939. It is thus interesting that he felt that the type specimens would have been safer in London—the obvious target of bombing by the German Luftwaffe—than in Singapore, which at the end of 1939, was nowhere near being embroiled in any conflict. To safeguard these type specimens, he arranged to have them permanently transferred to the British Museum (Natural History) in London but failed to get this done before the Japanese invaded Singapore. This transfer was not effected till after the war in 1946. We have no reports of activities of the museum after 1940.

When the Japanese attacked Malaya in December 1941, Chasen had the type specimens removed to the herbarium at the Botanic Gardens as the museum's proximity to the military headquarters on Fort Canning would make it a likely target of bombing. The specimens remained at the Botanic Gardens until after the British surrender, and E. J. H. Corner, Assistant Director of the Gardens, would then return whatever he could find to the Raffles Museum, where they remained till the end of the war.

It may well be that Chasen, who tried to escape Singapore before it fell to the Japanese, had also tried spiriting some of the more valuable type specimens with him. If that were so, those would also be lost for all time since Chasen's ill-fated ship, the *HMS Giang Bee*, was sunk in the Banka Strait in the Java Sea by a Japanese destroyer on 13 February 1942, the day before Singapore fell to the Japanese. All his possessions, including his notes and documents—and perhaps his type specimens—went down with the vessel.[59] With Chasen's death the history of natural history in Singapore passed on to a new phase and to a new generation of curators and scholars.

CHAPTER FIVE

The War Years
1942–45

WHEN EUROPE WENT to war in 1939, everyone had hoped that it would be brief and contained. It was neither. Japanese aggression in the East began with its conquest of Manchuria in 1937. And since Japan, Germany, and Italy had been enjoined as allies in 1937 through a series of treaties that integrated their military ambitions, it was only a matter of time before Asia would be engulfed by war. The Pacific War officially began with Japan's simultaneous attack on Pearl Harbour in Hawaii and on Thailand, Malaya and Hong Kong between 7 and 8 December 1941. In Malaya, the Japanese first landed troops in Kota Baru, on the northeast coast of the Malay Peninsula, as well as in Patani and Songkla in southern Thailand. The troops then proceeded southwards towards Singapore, reaching its shores in just 70 days.

From Singapore to Syonan-To

Lieutenant-General Arthur Percival, General Officer Commanding (Malaya), surrendered to the Japanese on 15 February 1942, and with that began three-and-a-half years of Japanese military rule in Singapore. For the thousands who endured the privations and terror of

5.1 Japanese troops marching into Singapore city, February 1942

this time, it was at best a period to be forgotten. To this day, animosities and prejudices continue to run deep and it is in this context that we should understand the efforts of a few key personalities in their attempt to save and protect the museum and its contents. Much controversy surrounds the account of how things took place after Eldred John Henry Corner published his book *The Marquis* in 1981.[1]

Eldred John Henry Corner (1906–96) was born in London, the son of a surgeon. He was educated at Rugby School, where he won a scholarship to study botany at Sidney Sussex College in Cambridge. There, he studied under Arthur Harry Church, then Reader of Botany. Specialising in mycology (the study of mushrooms), Corner came to Singapore in 1929 just after his graduation and became Assistant Director of the Botanic Gardens, working under the redoubtable Richard Eric Holttum (1895–1990). Many years later, John Kavanagh Corner, Eldred Corner's estranged son, would critically re-examine his father's account of these events and publish what is probably the most thorough and complete account of those days.[2] One of the big controversies surrounding Corner's actions was whether he had become a Japanese collaborator.

The battle for Singapore did not affect the museum building greatly. It was noted that the only physical damage suffered was when some artillery shells 'knocked off parts of the Library up-stairs and the room adjoining the Society's room.'[3] And thanks to the intervention of Professor Tanakadate and Corner, looting was prevented and all specimens remained intact until a burglary occured on 6 September 1942.[4] After the place was cleaned up and tidied, the museum was formally re-opened to the public on 29 April 1942. At the fall of Singapore, four persons were living on the premises of the museum, including Dr Carl Alexander Gibson-Hill, the Acting Health Medical Officer who, in the days preceding the surrender, had been appointed by the Undersecretary Wilfred Arthur Ward to take charge of the museum and the library.[5] Unfortunately, Gibson-Hill was unable to get along with Tanakadate and was sent to internment sometime around 20 February 1942.[6]

5.2 E. J. H. Corner with his son, John, mid-1941. Image from John Corner's My Father in His Suitcase (2013).

Quixotic Quest or Fool's Errand?

The Dramatis Personae

At the time of the surrender, Holttum was Director of the Gardens while Corner was his Assistant Director. Frederick Nutter Chasen was Director of the Raffles Museum and Library but had already perished the day before while trying to escape the island. The museum should have been in the hands of Chasen's next in command, Michael Wilmer Forbes Tweedie (1907–93), but he had been transferred to the Royal Air Force in 1941 as a Camouflage Officer, was interned once hostilities ceased, and then sent to Java as a prisoner-of-war. The museum was thus nominally left in the charge of its Archivist, Lim Soo Chye.

Corner's Dream and the Governor's Letter

On the day of the surrender, 14 February 1942, Corner had sought refuge at the Fullerton Building where the British Governor of the Straits Settlements, Sir Shenton Thomas, had set up a temporary office in the Singapore Club on the first floor of the building; he had moved there when Government House became uninhabitable. Also at Fullerton Building was William Birtwistle, Director of Fisheries for the FMS and the Straits Settlements, and soon to be an important figure in the events that ensued. That night, like so many others, Corner could not sleep with a rush of things racing through his mind:

> How that night of doubt and sorrow passed, I cannot recall. I began
> pondering by what means the treasure houses of knowledge which

had no military significance might be saved from what seemed surely to be a temporary occupation. If they were looted, I could not see that funds would ever come to start them again. Looting was the imminent danger. We had heard repeatedly during the past two months, as the withdrawal hastened to Singapore, how the looters broke into buildings before the Japanese had gained control, and how they smashed doors, cupboards, desks, safes, and boxes, scattering the contents, throwing books and files into heaps or bonfires, slashing pictures, and wrenching off taps to let the water run over books, papers, clothes, and carpets. I was thinking not just of the Botanical Gardens and the Raffles Museum, but of the libraries at Raffles College, the College of Medicine, Government offices, even at Government House, and the service they could render in the restoration of peace and order. Trophies would be carried off to Japan. The whole progress of science in the Malay Peninsula seemed doomed.[7]

The next day, Corner tried returning to his Assistant Director's house at the Botanic Gardens to collect his own manuscripts and drawings but was forced to turn back by Japanese sentries. While out and about, Corner heard that all British civilians were to be interned the next day. There was to be a parade at the cricket-ground (the Padang) at 11 am and all British persons were to assemble with their belongings and march to an unknown destination. That night, Corner could not sleep and kept 'thinking, thinking, and thinking, and fell asleep still thinking.'[8] The next morning, he woke up with a start,

> …laughing and exclaiming 'What a fool I am! Of course, of course!' In a lucid flash I had seen myself go to the Japanese authorities, explain the problem, and be received with open arms. There were no details, but a crowd of persons where I was happy.[9]

Corner dressed and sought out the Governor for his approval for his self-appointed mission. As he was headed downstairs, he saw the Governor's aide-de-camp who ordered him to report to the Governor who, coincidentally, 'had the same idea.'[10] According to Corner, the Governor 'wrote [him] a pencilled note in which he requested the Japanese authorities to preserve the scientific collections, libraries, and matters of historic interest, particularly at the Museum and Botanical Gardens,' and handed the note to Corner 'with a twinkle in his eye' and charged him to deliver it.[11]

Corner Meets the Japanese

By then, the Japanese had quickly established a *gunseikan-bu*, or military administration, under the command of Colonel Wataru Watanabe, and a new municipal government in the charge of Mayor Shigeo Odate and former Consul-General Kaoru Toyoda as his deputy.[12] On 17 February, the day British civilians had been told to assemble at the Padang, Corner locked himself up in his room, waited for the British civilians to be marched off, and then set off on his mission of delivering the Governor's note. He had heard that the Japanese headquarters was in the Colonial Secretary's office in the Municipal Building[13] and proceeded, with difficulty, to get there. There, he met Lazarus Rayman, hitherto President of the Municipal Commission, and showed Rayman the note. Rayman, who was there to hand over the municipality to Kaoru Toyoda, was also trying to keep certain British civilians out of internment in order that they could help repair and maintain essential services till Japanese replacements arrived. Rayman, realising the urgency of the matter, gave Corner precedence and ushered him to see Toyoda whom Corner remembered as a 'cultured and clever man.' Toyoda told Corner that a professor would be arriving from Saigon to take charge of scientific affairs and that he should return to the Municipal Building the next morning at 10 am and in the meantime remain in the Fullerton Building. Toyoda then provided Corner with an official pass and an armband to ensure his safety.[14]

'I Have Come Here in Order to Conserve the Cultural Heritage'

At Fullerton Building, Corner had met Birtwistle, Director of Fisheries, who had been ordered by the Governor to remain at his post on account of his familiarity with the Japanese fishermen whose boats he licenced.[15] Just minutes after Corner left the Municipal Building, Professor Hidezo Tanakadate,[16] the professor from Saigon referred to by Toyoda earlier, arrived at Toyoda's office. Tanakadate recorded this meeting in an article he sent to the Japanese daily *Asahi Shibun* in April 1942:

> In the morning of 17 February we travelled to Singapore City by car. We arrived at the city office and met people who worked there. They were chiefs of the military administration section, the

executive section, the industry section, and Mr Toyota, who was the ex-Consul-general of Singapore. When they saw me, they said. 'We have just received this from the ex-Governor', and handed me a letter. The letter said ·'We truly hope that you will conserve the Museum and the Botanical Gardens at least, and that you will engage Dr [sic] Corner, who was formerly the Assistant Director of the Botanical Gardens for that purpose.' I replied, 'I have come here in order to conserve the cultural heritage. The requisitioning of the Museum is my first objective.' At my words, everyone present, including the former mayor of Singapore, British and Japanese alike, glanced at one another and seemed pleased with my answer. Then they came to shake hands. It must have been a surprise for them to discover that a scientist from a university had come to conserve the local cultural heritage of an enemy country. The chief of the executive section said to me, 'I entrust you with all the cultural institutions' and asked, 'Will you accept the post of Director of the Botanical Gardens and Museum?' I answered 'Yes.'[17]

The next day, Corner joined Tanakadate to take over the museum and botanical gardens. Corner was also given the responsibility of increasing food production while Birtwistle was responsible for resuscitating the fishing industry. As a first step to conserving whatever was in place, Corner suggested to Tanakadate that notices be put on the doors of the museum, the library and the various offices in the Fullerton Building to prevent material from being looted or interfered with. Toyoda agreed and offered a small municipal car for Tanakadate's use.[18] Tanakadate proved to be extremely resolute in his effort to save whatever he could, even though Corner doubted if he had any authority to do what he was doing. Years later, Tanakadate sent shivers down Corner's back when he confided that when he was summoned to the Southern Army Headquarters in Saigon in February and ordered to go to Malaya, he was only 'treated as an Army employee with a rank of a sublieutenant but without remuneration as from February 15, 1942'. Indeed, Tanakadate later told Corner that his specific task was in fact to investigate 'useful minerals.'[19] As Corner recalled:

> ...It is now clear to me that, when we met and first spoke in the Municipal building, he perceived a greater responsibility much more in keeping with his character, changed his plans, and kept the original from my knowledge.[20]

Tanakadate's fearlessness appeared to have stemmed from two things. First, he knew that General Hideki Tojo, the Japanese Prime

Minister, had issued a direct order to all high commands 'that the contents of museums and libraries and all scientific collections were to be held and maintained in the countries where they belonged, for the benefit of the people.'[21] Second, Tanakadate had been at university with General Yamashita—the fearsome Tiger of Malaya—and actually called on him to support his plan. According to Tanakadate, Yamashita not only agreed but 'sought to promote his plans.'[22] In his report to Lieutenant-Colonel Archey—the officer in charge of Monuments, Fine Arts and Archives under the British Military Administration in 1945—Corner wrote:

> Professor Tanakadate was able to achieve this high-handed action
> (even to the extent of removing Nippon Military Notices from the
> Botanic Gardens Office), through his personal friendship with
> General Yamashita and, particularly, through the prestige of his
> father (or father-in-law) Baron Tanakadate, of the House of Peers
> in Japan.[23]

After securing the Museum, Takanadate secured the Botanic Gardens and, with Holttum's help, kept it going. In December 1942, Professor Kwan Koriba (1882–1957) of Kyoto University arrived to become Director of the Gardens and 'thereafter Mr Holttum ceased to have responsibility for garden work' even though he and Corner 'remained in constant touch

5.3 Japanese staff of the Syonan Museum, 1943. Seated (L-R): Yata Haneda (Assistant Director); Marquis Yoshichika Tokugawa; Yoshii (Secretary); Standing (L-R): Secretaries to Tokugawa— Sugawara, Omori, Tscuchida and Ishii

with the local staff and advised them in their work.'[24] Tanakadate also enlisted the help of Quan Ah Gun, the Botanical Gardens' Chief Clerk to act as the Chief Clerk of his new establishment, the Museum, Library and Botanical Gardens.[25] Tanakadate spent only 13 and a half months in Singapore but his actions in protecting Singapore's cultural properties ensured that the ravages of war did not obliterate a people's past. With the help of Corner and Birtwistle, he secured public and private libraries and prevented the destruction or theft of at least 80,000 volumes; he had Raffles' statue moved from its pedestal in front of the Victoria Memorial Hall to the museum for safekeeping, where it was prominently displayed in the entrance lobby; and he had Corner and Birtwistle sequestered many valuable British documents—such as Raffles' papers and manuscripts—in the rafters of the museum so that the Japanese underlings could not get their hands on them. Tanakadate was told to return to Japan on account of his perceived 'pro-British' leanings; he left in June 1943.

The Marquis (1942–44)

In September 1942, Yoshichika Tokugawa (1886–1976)—or the Marquis[26]—became the President of the renamed Syonan-to Museum and Botanical Gardens. He was the adopted son of Yoshiakira Tokugawa, the last *daimyo* of the Owari branch of the Tokugawa clan. An indifferent student right through his university years at the Tokyo Imperial University, he found his calling in the study of biology and botany. Prior to coming to Malaya, he had established the Tokugawa Institute for the History of Forestry (1923); the Tokugawa Reimekai Foundation (1931); and the Tokugawa Art Museum in Nagoya (1935). At the time of his appointment in Singapore, Tokugawa was Supreme Consulting Adviser to the Nippon Military Administration and Civil Governor of Malaya. By this time, the museum and gardens had been reorganised as a sub-department of the Department of Education, which came under the control of the Municipality.

During the Marquis' two-year tenure as President, he cleared up the museum and established an office for himself. The research facilities at the museum and library continued to attract some Japanese scholars but the public did not have access to these facilities. He studied the history of Malaya and Malay manuscripts at his office and also employed several Malays to help with translating the Jawi text into romanised Malay so he could read them. Furthermore, he oversaw the work of three Japanese Professors—Tanakadate and Yata Haneda (1907–95), at the Museum; and Kwan Koriba at the Gardens.

Two major publications—in English—emerged during the war years. The first was Michael Tweedie's *Poisonous Animals of Malaya* (1941), which had been in press when the war broke out and was thus published while Takanadate was in charge. Corner, with the help of Lieutenant (Dr) Tadamichi Koga (1903–86)—Director of the Ueno Zoo and attached to the military command—found many unbound copies from the broken-into and looted premise of the Malayan Publishing House. Koga found more copies in Kuala Lumpur and had them all bound at his own expense; he sent 500 copies to the museum.[27] The other was Colin Fraser Symington's *Forester's Manual of Dipterocarps* (1943),[28] which had a much more difficult gestation. Symington was Forest Botanist at the Forest Research Institute in Kepong, near Kuala Lumpur. He had spent years compiling and writing up the manuscript for the *Manual* and in November 1941 sent it to the Caxton Press for typesetting. Symington, who had escaped Singapore, but committed suicide in West Africa in 1943, never saw the final publication. When Corner told Tanakadate this story, the latter was deeply moved and said, 'But…this is a scientific work and must be printed.'[29]

Upon investigation, it was discovered that two persons who had worked closely with Symington were still in Singapore: Lieutenant Harold Ernest Desch and Charles Grenier, the manager of Caxton Press. Unfortunately they were interned. Tanakadate discussed

5.4 From left: The Marquis Yoshuchika Tokugawa, unidentified, Hidezo Tanakadate, and Corner. Note his seating position with head lower than Japanese 'seniors'. Image from John Corner's My Father in His Suitcase (2013).

Publishing Science in Changi Prison

A remarkable achievement during the war was the publication of *An Introduction to Malayan Birds* (1943) by Guy Charles Madoc (1911–99), at the prisoner of war camp in Changi. The son of an ornithologist in the United Kingdom, Madoc grew up with a love for birds and a keen eye for observing their behaviour. He was eventually posted to Malaya as a police officer. Having served both the British and Malayan governments in the intelligence and security areas, Madoc was interned as a civilian and had only 'a small fraction' of his bird notes with him (the rest were safely hidden at the Raffles Museum). To write his book, he relied largely on his memory and those of fellow internees who had formed an 'ornithological study group'.

> Within these four walls so much interest has been displayed in Malayan birds that I have felt encouraged to write this little book. Though I hope that the reader may find some matters of interest in it, a desire to enlighten my fellow-prisoners has not been the sole incentive to authorship. Its composition has enabled me to jot down many facts of which I no longer possess written record, and to crystallize and sift my knowledge.
>
> —Guy Charles Madoc, preface to *An Introduction to Malayan Birds* (1943)

Madoc wrote most of the book, with a borrowed typewriter at that, while other internees provided leather and paper stolen from the Japanese to make the book, and also helped to bind the book. Dr David Molesworth provided the illustrations, and Carl Gibson-Hill, who later became director of the Raffles Museum, helped in compiling the two appendices. The first edition was a single copy published in May 1943, and was widely circulated in the prison camp. Despite the conditions under which the book was produced, it is still considered a classic of Southeast Asian ornithology today.

Madoc kept extensive bird-watching notes, some of which he sent to the Raffles Museum. After his death, his daughter donated eight volumes of these notes to the British Museum (Natural History); they continue to be an important reference for ornithologists today.

5.5 *(Top) Madoc scrutinising a brown booby, one of his favourite bird species, on Pulau Perak, a Malaysian island*

5.6 *(Left) Madoc with some locals at the beach*

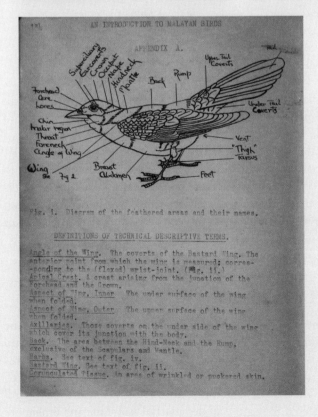

Plates and a diagram from Guy Charles Madoc's An Introduction to Malayan Birds *(1943)*

5.7 (Top Left) Black-Necked Tailorbird, Orthotomus atrogularis *(Temminck, 1836)*
5.8 (Top Right) Black-and-Red Broadbill, Cymbirhynchus macrorhynchos *(Gmelin, 1788)*

5.9 (Left) Anatomy of a bird

this with the Marquis, General Yamashita and Count Terauchi, and they decided that the two men should be asked to help prepare the manuscript, if it could be found. Corner had learnt from Symington's wife that there should be a copy of the typescript at the Caxton Press in Kuala Lumpur. Tanakadate decided to go in search of the typescript personally and drove Densch in his car to Kuala Lumpur, where they recovered the galley proofs from the Caxton Press office. Describing the extent to which these 'Japanese men of science' went to get the publication out, Corner wrote:

> It was decided to publish the work (500 copies) on the ground that it would be more likely to survive the War in that way than as a single galley-proof, for the whereabouts of Mr Symington and his manuscript were unknown. The cost of printing was met personally by Prof Tanakadate and by Marquis Yositika Tokugawa, who acted as president of the Museum and Library. It was insisted by Prof Tanakadate that the book should conform exactly with the previous series of *Malayan Forest Records*, of which it is No 16, so that it should stand the test of time, as a scientific work, regardless of hostilities and racial prejudice. He there added a brief preface, as a single page of romanised Japanese, and he issued the Manual from the Museum to give it official standing and to prevent pilfering of the stock by what he called 'common people.'[30]

Two other scientists assisted the Marquis. At the gardens, he had Kwan Koriba, a famous botanist who had recently retired as Professor of Botany at the University of Kyoto, and at the museum, Yata Haneda, a medical doctor who went on to become one of the world's authorities on luminescent organisms. After the war, Haneda returned to Japan to become Director of the Yokosuka City Museum. When the Marquis retired at the end of 1944, Haneda was appointed Director of the Museum and Gardens while Koriba was appointed Director of the Gardens. They remained in office till the end of the war.

Picking Up the Pieces

The suddenness of the Japanese surrender in August 1945 caught the British off guard as they had expected to fight their way back into Singapore and return as 'heroes'. Instead, they found themselves scrambling to take control of Singapore and secure its key institutions. The British Military Administration (BMA) took control and administered Singapore under martial law. Entering Singapore along with the reoccupying forces was Lieutenant-Colonel Gilbert Archey (1890–1974), Officer-in-Charge in the Monuments, Fine Arts and Archives (MFAA) division—a 'Monuments Man.'[31] He was charged with taking over the museum and library and conducting an investigation into the extent of damages or loss they suffered during the Japanese Occupation. Corner and Birtwistle were asked to provide a report to Archey, as was Henry Maurice Pendlebury, the last Director of the FMS Museums.

5.10 Gilbert
Archey, the
'Monuments Man'

Before the war, Archey had been Director of the Canterbury Museum in New Zealand and later the Auckland Institute and Museum. Born in York, England, he and his family immigrated to New Zealand in 1892. He was educated at Canterbury College where he graduated with an MA in Zoology in 1913. During World War I, Archey had seen action as a member of the New Zealand Field Artillery and held the rank of Captain. He was now seconded to the BMA as MFAA officer with the rank of Lieutenant-Colonel, with the responsibility to protect religious buildings, monuments, museums and archives in Malaya, Thailand, Java and Indo-China.

As a precaution, all European personnel who were not interned at the point of the British return were taken into custody. Corner was only released at 4 pm on 6 September 1942 and 'began the work of reorganisation at the Museum.' He handed the museum over to Archey on 8 September, the day before Birtwistle was also released.[32] In his comprehensive report to Archey, Corner recorded all the 'losses' suffered by the museum during the Occupation. The list is extremely short and with respect to the zoological collection, only some duplicate skins of birds, shells and corals had been removed by Dr Koga for presentation to Emperor Hirohito.[33]

In addition to answering the various queries regarding the museum, Corner and Birtwistle also helped in 'clearing' the Japanese scientists still on the island. In accordance with Corner's explanations, Archey made the following recommendation to BMA Headquarters:

> These officers were assiduous protectors of the Museum, Library and Gardens. They prevented Military occupation of the premises, they maintained the collections and books with the exception of certain duplicates of which they gave records, and gave effect to suggestions made by Mr Corner and Mr Birtwistle for bringing to the Museum for safety such books, archives and records as could be secured. Tribute must be paid to the enlightened attitude of these Japanese officers, by whose action the Museum, Library and Herbarium have been preserved. Marquis Tokugawa and Prof Tanakadate are in Japan; Dr Koriba and Dr Haneda are prisoners in

Malaya and it is suggested that consideration be given to the above
facts when disposal of enemy prisoners is being decided.[34]

To this end, both Koriba and Haneda were given access to the library
of the botanic gardens and the museum to conduct their research, and
Haneda—for whom Corner wrote a long testimonial—was allowed to
retain his papers for publication when he returned to Japan.[35]

The most immediate task for Archey and the BMA was to find
suitable replacements for personnel lost due to the war. Holttum had
returned to England[36] and Archey considered retaining Corner as
Acting Director of the Gardens the best option. The main problem was
the museum. Henry Maurice Pendlebury, Director of Museums, FMS,
was in such poor health that his return was unlikely. Archey proposed
inviting applicants for the post of Director of Raffles Museum and
Library on condition that the existing staff, namely Michael F. W.
Tweedie and Hubert Dennis Collings—who were prisoners-of-war—
had the right to put in their applications as well.[37] In October 1945,
when Pendlebury died, Major-General Ralph Hone, Chief Civil Affairs
Officer for Malaya suggested that Archey be prepared to temporarily
act as Director until Tweedie or Collings or both of them returned.[38]
Unfortunately, this was not possible as Archey was unable to extend his
leave any further: his employers, the Auckland Institute and Museum,
had made a number of plans on the assumption of his timely return to
New Zealand.[39]

Archey completed his tour of duty in March 1946[40] but there is little
indication of who took over the musuem at this point. We do know that
Tweedie returned to Singapore in June 1946[41] and assumed the post of
Director of the Raffles Museum and Library. During the war, he had
been taken to Java in March 1942 as a prisoner-of-war and interned at
the Boei Glodok Prison Camp.[42] With Tweedie back as Director, and
losses at the museum fortuitously limited by the actions of Corner,
Birtwistle and the Japanese directors, it would just be a matter of time
before the museum returned to its former glory.

In this regard, the museums in British Malaya survived the
war fairly intact. Save for the American bombing that accidentally
destroyed one wing of the Selangor Museum in March 1945, the
museums of Perak, Sarawak and Singapore survived almost in their
entirety. In comparison, the American bombing raids to liberate
Manila almost completely destroyed the Manila Museum of Natural
History. Interestingly, the destruction of cultural property in Asia
generally, and of museums in particular, resulted more from Allied
efforts to retake their former territories than wanton destruction by
the occupying Japanese forces.

Re-opening of the Museum

On 23 October, the Raffles Museum was reopened to the public. Below is the first post-war touristic description, prepared by Corner:

The Museum building was opened on 12th October, 1887 by H. E. the Governor Sir FA Weld in the Jubilee year of Queen Victoria. Since then it has been closed to the public only on the two occasions of the Japanese occupation (15th February–26th April, 1942) and of the Japanese surrender (20th August–11th September, 1945).

The object of the Museum is to display the culture and livelihood of the human races inhabiting the truly Malayan region of Sumatra, Borneo, Java and the Malay Peninsula, and also the zoological natural history of this region. The aborigines of the Malay Peninsula are the jungle-people called the Negrito and Sakai, and all others are immigrants from the surrounding countries, many of them within historic times. The animals on the other hand are the true denizens of the evergreen tropical rain-forest which originally covered the whole peninsula.

The central hall now holds the statue of Sir Stamford Raffles which was brought to the Museum for safety during the Japanese occupation. Behind it is a case showing some of his letters and early history. On the walls are paintings of the early European settlements of Penang, Malacca and Singapore.

Immediately on the left of the entrance is the sole fragment of the great stone with indecipherable inscription which stood at the mouth of the Singapore River when Raffles first landed.

In the left wing are the show-cases illustrating the crafts, games and ceremonies of the people of Malaya. In the right wing are displayed models of Malayan sea-craft and casts of Malayan tombstones from the inscriptions on which much historic information has been gathered.

Upstairs the right wing contains the show-cases of Malayan weapons, silver and sarongs, together with a small collection of old coins. Adjoining the silver-room is a new extension for the subject of prehistory, in which are displayed the local flint implements.

The rear part of the building, upstairs, contains the natural history specimens which are arranged in zoological groups beginning with the Mammals on the left and leading through the Fish room to the main gallery with Reptiles, Birds, Insects, Crustacea, Corals and Sponges. The newest exhibits are the casts of local fish and of local Crustacea (on the extreme right), the former being of particular interest in illustrating the variety of the local

5.11 The original display of mounted fish specimens in the Raffles Museum, circa 1931. Several of the large specimens are still in the museum today.

fish specimens. The collection of insects is small because their study was centred in the Selangor Museum, Kuala Lumpur.

Among the Bird exhibits, that showing the nesting habit of the Malayan Hornbills is noteworthy.

The main collections of bird and mammal skins, together with the zoological library, are upstairs on the left, or Director's wing.

The Museum has suffered very little damage or loss through the Japanese occupation.

Remarkably, the museum's war-time losses were few. Even so, not all specimens survived. In May 1946, the Director of the British Museum submitted a list of the type specimens thought to be in Singapore. A search was made and it was discovered that 71 of the 103 bird types and 99 of the 113 mammal types were indeed in the Raffles Museum and were, accordingly sent to London under the original agreement with Chasen. Of the original list, 32 bird types and 14 mammal types went missing.[43] Gibson-Hill, who was appointed Curator of Raffles Museum in 1947 speculated:

> No trace could be found in Singapore of the type specimens of the remaining 32 birds and 14 mammals. Enquiries were sent in April

to the Museums of Leiden, Amsterdam, Buitenzorg and Kuching, and they do not appear to be in any of these collections. Where this is not so, they must have been lost during the moves to and from the Botanic Gardens in 1942, or have been amongst the material which disappeared when Chasen's house was looted five days before the fall of Singapore. In either of the latter cases, they must be considered lost for all time.[44]

Foenander's Seladangs Go to War

Eric Carl Foenander of the Forestry Department in Perak was one of Malaya's most famous big-game hunters and he had a large number of trophies to show for it. Foenander, who was of a Ceylonese Eurasian, even wrote a book entitled *Big Game of Malaya: Their Types, Distribution and Habits* (1952). In January 1940, Frederick N. Chasen, Director of the Raffles Museum and Library, visited the Selangor Museum and 'saw some beautiful seladang heads' that had been deposited by Feonander there.[45] Constantly on the lookout for beautiful display specimens, Chasen wrote to Foenander, informing him that if he ever felt 'inclined to dispose of any of these heads, they would be most acceptable to the collections in this museum where there are only two old and not very good heads on exhibition.'

Foenander, quite clearly pleased with Chasen's compliments, replied the next day, informing him that while he was 'reluctant to dispose of the heads', he was happy to loan them to the Raffles Museum for five years (with an option for a further five years after that) for display. Foenander also offered Chasen six pairs of elephant tusks. Chasen was delighted and accepted the offer, promising to exhibit them with full attribution of its provenance to

5.12 The seladang, Malaya's largest land animal after the elephant, as photographed by E. C. Foenander.

Foenander.[46] On 11 March 1940, four seladang heads and two elephant skulls and a spare lower jaw arrived in Singapore. They then went on display and the elephant tusks were considered the 'Most Important Acquisition for Many Years.'[47]

Shortly after the Japanese invasion of Singapore, Foenander became worried about the condition of his specimens and wrote to the Director of the Museum in Syonan-to to ask if his loan exhibits were 'still safe and intact.'[48] He received a reply from Hidezo Tanakadate, informing him that the Raffles Museum no longer existed but was now known as the Syonan Museum. He was also told that no exhibits may be removed from the museum 'By Order', and that in any case, as Foenander had dealt directly with Chasen, it was Chasen who was responsible for the exhibits and Tanakadate could not accede to his request.[49]

Anxious not to upset the Japanese, Foenander wrote back apologetically, assuring Tanakadate that he had 'no desire to have them returned…at this time' and that 'it would be very selfish to ask for their return at this juncture and so deprive the public of a chance to view good specimens of the heads of Malayan types of these animals.' His enquiry, he assured Tanakadate, 'was prompted solely by sentimental reasons' and that 'through the foresight and generosity of our Government, these trophies are now ensconced in the safest place in Malaya—the Syonan Museum.'[50]

After the Japanese surrender, Foenander decided to enquire once again. This time he wrote to the Director of the Raffles Museum and received a most comforting note from Lieutenant-Colonel Archey, who told him that his trophies were all there and in order, and that his offer of an extension of the loan for exhibit would be gladly received by the Raffles Museum Committee 'when reconstituted.'[51] In February 1946, Archey wrote to Foenander, accepting his offer to have the trophies on display at the Raffles Museum for another two-year period. Foenander tried to personally call on the Director of the Museum on 3 September 1947 but did not manage to meet Tweedie. Instead, he found Gibson-Hill, the Assistant Curator of the Museum to whom he enquired about his trophies. Foenander could find no sign of them. He was infuriated at having to look around for them and found them in a store room in shocking conditions:

> I regret to say that I was shocked at the condition of my valuable and coveted Seladang trophies. The fine chengal timber shields with brass plates on which they were mounted had been removed and could not be traced. Lying as they were on the floor without protection and constant care the trophies were exposed to attack by the larvae of moths and the two best heads show disfiguring bald patches due to the hair having been bitten off. There is also evidence of careless handling in their removal to the store-room.[52]

Foenander then indignantly requested the return of the heads immediately. Gibson-Hill replied, explaining that it was Chasen who ordered the heads removed from their mounting shields and moved to the storeroom just before the war to prevent them from being damaged or looted and that Tweedie had ordered the removal of the elephant tusks and skulls from display for the same reasons. At the same time, Gibson-Hill regretted the deterioration in the specimens as there was no taxidermist at the museum since 1942.[53] This appeared to mollify Foenander, who wrote back to express his gratitude in Chasen's decision to store the Seladang heads in the storeroom before the war. And in view of the fact that was no taxidermist or executive staff from 1942 to 1946, Foenander conceded that the 'damage to these Seladang heads [was] remarkably light'.[54] The heads and tusks were returned to Foenander on 27 November 1947.

CHAPTER SIX

Becoming the National Museum
1946–70

WHEN CIVIL GOVERNMENT was restored in 1946, the old Straits Settlements was disbanded and Singapore became a separate Crown Colony. Penang and Malacca were, together with the Federated Malay States and Unfederated Malay States, merged into a federation known as the Malayan Union. And while the British did not return quite as triumphantly as they had wished, many Singaporeans were at least relieved to see them back—such were the privations of the Japanese Occupation. Fortunately, Singapore bounced back fairly quickly, given the worldwide demand for rubber and tin. However, poverty was endemic in the city area and housing was desperately short.

Back on Its Feet (1945–50s)

At the museum, things were beginning to go back to normal, especially after the return of Michael Tweedie as Director. Tweedie, who had been interned in Java during the war, got down to work immediately.[1] In September 1948, in response to the proposal to establish a full-fledged university in Singapore, he presented a proposal to have the university 'affiliated with the Raffles Museum and Library and so situated in Singapore.'[2] Tweedie saw many benefits of such a close affiliation and argued, especially with regard to the Zoological Collection, that as the museum functioned 'as the headquarters of the Zoological Survey of

Malaya,' and because at least two of its officers 'have been and will be zoologists', it has built up 'a large reference collection of vertebrate and invertebrate animals.' The University of Malaya would eventually be established in 1949 in Bukit Timah at the former campus of Raffles College (Malaya's first tertiary institution that had been established in 1928). The 'close affiliation' between the collection and the university that Tweedie envisioned would not materialise until many years later.

Tweedie and his staff spent time inventorising the museum and making lists of lost specimens and artefacts. In December 1946, Tweedie was informed by H. C. Lepper, Assistant Economic Adviser to the United Kingdom's Liaison Mission in Japan, that its Looted Property section had informed him that 185 stuffed birds 'had been found in Japan which were said to have been looted from the Raffles Museum, Singapore.'[4] In November 1947, these were shipped back to Singapore in three crates. Two crates, containing 164 specimens, arrived safely at the museum but the third went missing. It was later found to have been mistakenly shipped to London and was redirected to Singapore.[5] Tweedie also visited Sarawak in 1947 and added collections of Malaysian Crustacea and Land Mollusca to the Singapore collection. These included some land molluscs presented by the curator of Sarawak Museum, Edward Banks. C. S. Ogilvie, Warden of the King George V National Park in Pahang and staunch supporter of the museum, sent a collection of freshwater fish from his park.[6] He also regularly helped the museum identify many of the fish in its collection with corresponding Malay names.

Tweedie went on long leave from the end of 1947 to April 1948, leaving C. A. Gibson-Hill, the Assistant Director, to act as Director. In the meantime, the museum's ethnographical assistant, Abu Bakar bin Pawanchee, sailed to England where he was to study anthropology for a year at Cambridge on a British Council Scholarship.[7]

1948 saw multiple changes for the museum. First, the museum and library buildings were renovated and painted white and green on the outside and cream, white and green on the inside.[8] Secondly, improvements were made to the Zoological Galleries by the installation of 'informative labels, giving a short account of the habits, distribution and other interesting facts concerning the animals exhibited' and photographs and diagrams were included where appropriate. As a result of these new labels, the number of exhibits had to be reduced and specimens from Sumatra, Borneo and Java were 'removed to the reference collections and the exhibits confined almost entirely to the Malayan fauna.'[9] Lastly, the museum began actively collecting specimens for its Reference Collection again. Like in previous decades, the museum also relied on its supporters and well-wishers to present and name specimens. Work on the cataloguing of the Reference Collection also commenced in earnest, starting with the Bird Reference Collection, which had some 29,000 skins.

Sometime before 1949, the posts of Curator and Assistant Curator were redesignated to reflect the importance of the museum's ethnological and zoological collections. There would henceforth be a Curator of Anthropology and a Curator of Zoology. Hubert Dennis Collings, the archaeologist who joined as Assistant Curator in 1934, became Curator of Anthropology while Gibson-Hill was made Curator of Zoology. Much time was spent relabelling and rearranging the mammals and birds exhibits. Many of the mounted exhibits were faded and old and needed replacement and while replacement specimens could be found, the lack of a taxidermist meant that some of the specimens had to be sent to London for mounting.

The next decade was to be a period of calm and transition for the museum. With Tweedie at the helm and Gibson-Hill as his deputy, there was a sense of continuity as both men had been involved with the museum since before the war. It was a return to the pre-war routine of relabelling, rearranging and collecting. It seemed like nothing much had really changed. However, the situation was quite different outside the museum.

Major political changes were afoot as Singapore lurched nervously towards self-government and eventual independence from Britain. While pro-independence forces emerged in the immediate aftermath of the Japanese Occupation, it was not till 1953 that mass-based organisations began agitating for greater autonomy and self-government. Such pressures came primarily from the trade unions and nascent political parties. With the Malayan Communist Party (MCP) proscribed and outlawed, the contest for supremacy of the left outside the MCP was to see the emergence of the People's Action Party in 1954 and the Barisan Sosialis (Socialist Front) in 1961. The rising tide of nationalism and Britain's commitment to withdraw from Singapore would mean an urgent need to find local replacements for expatriate staff at all levels.

Staff Changes & the First Local Curators

In 1950, the entire staff of the Raffles Museum and Library—except for the daily-rated staff—was placed on the permanent establishment. That same year, Mohamed Shareff bin Hashim was promoted to the post of Taxidermist, a post that had not been filled since 1942.[22] The post of Curator of Anthropology was not filled till December 1952 when Prince John Lowenstein was appointed. In the interim, Margery Topham was engaged as a temporary officer.

Michael Wilmer Forbes Tweedie (1907–93)

Michael Wilmer Forbes Tweedie was born on 2 September 1907, the son of Maurice Carmichael Tweedie and Mildred Le Gros Clarke.[10] He was educated at Cambridge University where he read natural sciences and specialised in botany, zoology and geology. Upon graduation in 1929, he was employed by the Shell Oil Company in Venezuela as a paleontologist.[11]

He joined the Raffles Museum as Assistant Curator in 1932. Recalling his early days at the museum, Tweedie tells us that back then leaders of the museum like Kloss (who just retired when he joined), and Chasen 'interested themselves solely in mammals and birds' and that as Smedley—who retired at the same time as Kloss—had hitherto concerned himself with reptiles, he was then placed in charge of the reptiles, fish and invertebrates.[12] Although Tweedie's 'favourite animals were always the crabs,'[13] his interests were wide-ranging and he studied and published prolifically in many areas, including herpetology—his *The Snakes of Malaya* (1953) was a bestseller and continues to be a standard reference for all Malaysian and Singaporean herpetologists.[14] Unlike

6.1 Tweedie at the Raffles Museum

many zoologists of the old school—like Robinson and Kloss—Tweedie was not possessive about the specimens he collected. For him, science and knowledge mattered above all and he was not too concerned with whether he was the first to find and name a new genus or species:

> The fame of Tweedie however, laid not only in his scientific publications. Tweedie had the foresight (for which many of us today are grateful) to send material, including his beloved crabs, to other taxonomists who were more adept in certain areas than

6.2 Tweedie visiting the Zoological Reference Collection in 1988 with Professor Lam, Head of the Zoology Department

he was. Many of these specimens were collected by himself during his many travels. Our knowledge of the Brachyuran and Anomuran crab fauna of the Sunda Shelf and Indo-West Pacific owes a great deal to Tweedie's endeavours. In gratitude, these carcinologists have honoured Tweedie by naming many species after him.[15]

Tweedie himself explained his collection strategy as such:

Brought back to the museum in tanks of alcohol, the specimens were sorted into categories and a search made in the zoological journals for the names of suitable specialist taxonomists, mostly working in Europe and North America. A letter would be sent to such a specialist asking him or her whether a collection for study would be acceptable and offering the retention of duplicate specimens and publication of the report on the collection, in English, French or German, in the *Bulletin of the Raffles Museum*. This sort of approach was usually well received, and papers were published on Malayan planarian worms, earthworms, leeches, scorpions, pseudoscorpions, Opiliones, Pedepalpi, Isopods, centipedes, molluscs, various crustacea and fishes.[16]

Even after his retirement from the museum in 1956, Tweedie continued to be active, maintaining contact with the museum and the Reference Collection, as well as with scientists and naturalists around the world. In 1988, aged 81, he even travelled back to Singapore to attend the opening of the Zoological Reference Collection's facilities at the National University of Singapore. He died in 1993.

In 1955, when Singapore became a self-governing colony under the Rendel Constitution, a Malayanisation scheme provided scholarships for locals to be trained to fill key administrative and management posts in government and government agencies. The same year, a Departmental Scholarship was made available for a graduate in zoology to receive a year's training for candidates for the posts of Curator of Zoology and Curator of Anthropology. It appears there were no successful scholarship candidates in 1955 and 1956.

In 1957, Eric Ronald Alfred (b. 1931), a local zoology graduate from the University of Malaya in Singapore, joined the museum as Curator of Zoology. Of Indian or Ceylonese extraction and born in Batu Pahat, Johor, he thus became the first local person to hold the post of Curator. Alfred came to Singapore in 1949 to study zoology at the University of Malaya. His first contact with the museum was during the course of his studies when his lecturer John Roscoe Hendrickson (1921–2002) sent him to the museum to have some species identified. There he met Tweedie, Gibson-Hill, and other staff whom he recalled as being extremely helpful.[23] When he saw the marvellous library, his 'eyes popped out.' Leafing through the first edition of Pieter Bleeker's *Atlas Ichtylogigue* (1862–77), with its stunning colour illustrations for the first time, he decided that the museum would be a wonderful place to work.

Carl Alexander Gibson-Hill (1911–63)[17]

Carl Alexander Gibson-Hill must surely be the most versatile scholar to helm the museum. He was born the son of an engineer on 23 October 1911 in the suburbs of Newcastle-upon-Tyne. A precocious talent blessed with a pathological curiosity, Gibson-Hill learnt carpentry and photography from his father and excelled in both. He even built his own display cabinets for his personal collection of natural objects that he collected in his neighbourhood. As a teenager, he went around old churches and monuments, taking rubbings and documenting them. In 1936, using the pseudonym John Lisle, he published his first publication *Warwickshire*, a 300-page tome on the sights, monuments and natural history of the countryside where he lived.[18]

Gibson-Hill studied at Malvern College where, kept indoors by the inclement weather

6.3 *Display of Christmas Island birds that Gibson-Hill helped assemble*

and strict school rules, he developed his talents in drawing, bird-watching and butterfly collecting. By the time he graduated, he won the Gale Prize for his knowledge of entomology, and the Boldero Prize three times for his knowledge of natural history. He then went to Pembroke College, Cambridge, where he studied natural sciences, graduating with a second in the Natural Science Tripos in 1933. To please his parents, Gibson-Hill enrolled at the King's College Hospital Medical School to study medicine, graduating in 1938. There he met and married a fellow houseman, Margaret Halliday, and soon after their wedding, departed for Christmas Island where he served as Resident Medical Officer at the Christmas Island Phosphate Company.

To get to Christmas Island, Gibson-Hill had to travel through Persia, Afghanistan and Cambodia before arriving in Singapore. There, he met F. N. Chasen, Director of the Raffles Museum. Chasen, who had gone on a collecting expedition to Christmas Island in 1932, instructed Gibson-Hill to help build on the existing bird collection. By the time Gibson-Hill completed his stint on the Island in 1940, he had collected some 200 bird specimens. From Christmas Island, Gibson-Hill proceeded to the Cocos-Keeling Islands where he was joined by his wife, who had completed her medical residency. They spent ten and a half months there exploring the islands and getting to know John Sidney Clunies-Ross, the owner of the Islands. This friendship gave Gibson-Hill access to the Clunies-Ross family documents, which he would later draw on to write the history of the islands.

At the end of 1941, Gibson-Hill and his wife completed their posting and returned to Malaya. She found a job at the Alor Star Government Hospital in the state of Kedah while he was appointed Health Officer for the rural districts. Gibson-Hill returned to Singapore in December 1941, not knowing that it would fall to the Japanese less than two months later. Chasen had already left the museum and was preparing to evacuate and it is not clear under whose or what authority Wilfrid Arthur Ward, Under-Secretary for the Straits Settlements, proceeded to appoint Gibson-Hill as Assistant Curator and place him in charge of the museum, which was why Gibson-Hill was staying on the museum premises when E. J. H. Corner turned up with Professor Tanakadate to take over the premises on 15 February. As mentioned earlier, Gibson-Hill did not get along with Tanakadate and was, on 23 February, sent to Changi along with all other British residents.

In prison, Gibson-Hill remained active, mentally reviewing the draft publication of his island postings. He was also Secretary of the Leisure Hours Committee for internees and conducted lectures on Malaya alongside other scholars who were also interned. The leisure time afforded him while he was interned allowed him to indulge his passion for ornithology and art and he even illustrated the 1943 issue of Guy Charles Madoc's *An Introduction to*

6.4 Raffles Museum Director Gibson-Hill (left) showing Yang di-Pertuan Negara Yusof bin Ishak around a photography exhibition at the National Library Lecture Hall in 1961

Malayan Birds.[19] After Singapore's liberation, Gibson-Hill returned to England where he continued to work on two books on British birds, published later as *British Sea Birds* (1947) and *Birds of the Coast* (1948).

He returned to Singapore in 1947 and rejoined the museum as Assistant Curator of Zoology, acting as Director while Tweedie was on long leave from the end of 1947 to April 1948. At the same time, Gibson-Hill was truly indefatigable. From 1947 to 1949, he also served as Acting Professor of Biology at the College of Medicine. In October 1949, Gibson-Hill completed and published the *Annotated Checklist of Malayan Birds*, as No. 20. of the *Bulletin of the Raffles Museum*; a preliminary checklist of Malayan mammals; and a report on cetaceans for the Sarawak Museum. In 1950, Gibson-Hill became the most senior member of the staff after Tweedie when H. D. Collings—who had been on the staff since 1935—went on leave and then resigned. That year, he published what is probably his most outstanding natural history study 'Notes on the Sea Birds Breeding in Malayan Waters.'[20] On top of his research and publication work, he also made extensive studies on local sea-going boats.

When Tweedie went off on long leave in 1951, 1955 and 1956, it was Gibson-Hill who stood in as Director. On top of his museum duties, Gibson-Hill found time to play an active role in the Malayan Branch of the Royal Asiatic Society, editing its journal from 1948 to 1961. He became the Society's President in 1957 when Tweedie retired.

Gibson-Hill was the last Director of the Raffles Museum as it was then called. In 1960, the museum was renamed the National Museum and placed under the Ministry of Culture. He retired two years later in 1962, just past his 50th birthday. A year later, Gibson-Hill was found dead in his bath. In response to speculation as to whether Gibson-Hill had taken his own life, Eric Alfred, his successor at the museum, states:

> I'd rather think it is 50-50 because he was a diabetic, quite a bad diabetic. One day he came to see me and said, 'There's this new drug that's come out. It's called insulin. I'm going to try it.' That's when I knew of his diabetic condition. He passed away during his bath. He died of a diabetic coma. He didn't kill himself. But he used to consume mescaline tablets to stay up late. He used to pop this all the time because he wanted to stay awake and work. He really worked day and night. He was a brilliant man. He was also being persecuted by some people from the Ministry of Culture, trying to get him to retire on medical grounds because they didn't want an 'angmoh fella' in charge.[21]

It was most unfortunate that so sensitive and brilliant a soul as Gibson-Hill should become the last expatriate Director of the Museum at the time when the tide of nationalism had turned decidedly against expatriates. While Singapore had certainly seen more than its fair share of mediocre white men ('angmohs') strutting about and lording over the locals, Gibson-Hill was not one of them. That did not seem to make any difference to some of the new 'local men' who took over from the British expatriates and sought to undo more than a century of racism by persecuting the remaining British and European expatriates.

Tweedie encouraged him to apply for a job and 'sort of conducted an interview before the actual interview' with the Public Service Commission.[24] On 15 August 1957, Alfred joined as Curator of Zoology. A week later, Christopher Hooi, another local graduate, who studied history, joined the museum as Curator of Anthropology.[25] Hooi did not stay long; he later joined the Administrative Service and was posted to the Ministry of Culture. When Gibson-Hill retired as Director in 1962, Alfred succeeded him, taking over as Acting Director of the National Museum.

The Building and Collections

There was little physical change to the museum, largely because by 1956, the anticipated split between the library and the museum—so long advocated by a long line of directors—looked more like a reality, and the library would soon get its own home. In 1953 Lee Kong Chian, the well-known business tycoon and philanthropist, offered to donate a large sum of money towards the establishment of a public library. It was an offer that was too good to be true, but Lee had one condition: the library had to stock books in Asian languages as well as in English so as to be fully representative of the languages and cultures of Singapore's inhabitants. The Colonial Government agreed to two more conditions: they would meet any expenditure beyond the sum guaranteed by the Lee Foundation from public funds, and move the books at the present Raffles Library to the new building when ready. This would thus mean that the museum—for so long strapped for space—could now occupy the entire building. With the planned split in the mind, the museum created, in 1954, a separate Museum Library that would be a technical library 'consisting of books and journals concerning zoology, anthropology and archaeology'.[26]

6.5 Eric Alfred, Acting Director of the museum (1967–74), also its last Curator of Zoology (1956–74)

In the meantime, approval was given in 1954 for the exhibition galleries to be renovated and exhibits rearranged 'in accordance to modern display methods'.[27] Gibson-Hill, who was then in England, was assigned to 'consult a firm specialising in museum furniture'.[28] In furtherance of this scheme, a four-year plan to modernise the galleries was put in place and work commenced under Gibson-Hill's direction since it was he who drew up the initial plans for renovations the year before. The following year, construction work began on two rooms that housed the Archives and the Directors' Office. A mezzanine floor was built in the first room such that the upper space was made into a single room and the space below divided into two rooms, one of which would be a photographic dark room. The other room, which could be accessed by a gallery and stairway, was equipped with steel shelves from floor to ceiling to accommodate books. The whole space would be air-conditioned.[29]

In terms of acquisitions, two of the most interesting donations came in 1950. The first was an autograph of the great naturalist Alfred Russel Wallace, which was presented to the museum by Dr Frank Fortescue

Laidlaw (1876–1963), an expert on dragonflies and molluscs. The second was a huge cobra caught by the greensmen at the Island Golf Club in Thomson Road on 10 July 1950:

> The snake was brought alive to the Museum and proved to be a King Cobra or Hamadryad…After it had been chloroformed and killed it was found to measure 15 feet 7 inches in length and to weigh 26 ¹/₂ pounds. A plaster mould was immediately made of it from which a papier-mache cast has been prepared for the galleries, but painting of the cast was not completed by the end of the year. The skin has been preserved and the head presented to the Zoological Faculty of the University of Malaya to provide a skull for teaching purposes.[30]

The collection continued to be added to through collecting expeditions, and donations. One of the most significant was a 'very fine collection of butterflies from the Langkawi Islands' by M. J. V. Miller, a collector. The collection consisted of some 360 species, some of which were very rare, and almost all of which were 'in beautiful condition.'[31]

6.6 Visitor viewing
the butterfly
collection

Moving in Different Directions in the 1960s

The administration of the Raffles Library and Museum was eventually separated into two in 1957 through the passage of three bills in the Legislative Assembly.[32] Firstly, Ordinance No. 30, the Raffles Museum Ordinance, constituted the museum as a separate legal body and institution.[33] Then, the Raffles Library Ordinance turned Raffles Library into a free public and national library by abolishing subscriptions and renaming it the Raffles National Library.[34] In 1960, the museum ceased to be known as the Raffles Museum and was renamed the National Museum, it would now operate under the Ministry of Culture.[35]

The National Museum and Nation-Building

An educational and nation-building mission became increasingly important in the course of the 1960s as Singapore edged towards nationhood. This included a cultural and social education for citizenship, and an emphasis on science and technology for industrialisation and economic development. At the time of its creation, the National Museum inherited the entire collection of the old Raffles Museum, which included the ethnological collection and the natural history collection, as well as its general collection of art works and historical artefacts of Singapore and the region. However, the government had new plans for the museum. In his speech to the Legislative Assembly in 1960, the Yang di-Pertuan Negara, Yusof bin Ishak, briefly outlined the new proposed role of the National Museum:

> The National Museum will be reorganized to make it more representative of local customs and traditions. There will be more stress on the wealth of cultural material out of which we hope to weave a Malayan culture. It will emphasize more the history and contributions that the many races have made towards the creation of modern Singapore.[36]

With this message from Yusof Ishak, it was clear that a fundamental shift was about to occur. The museum, for so long a centre of scientific learning, research and education, was now to be mobilised for nation-building, which meant a corresponding shift away from the natural sciences towards history, culture and anthropology. With this object in mind, Alfred proposed creating the new post of Curator of Historical Studies. This would enable the burden of work to be distributed between the Curator of Anthropology—who would be entrusted with Material Culture, and Social and Physical Anthropology, as well as Prehistory and Archaeology—and the Curator of Historical Studies

6.7 *Students from Bukit Panjang Government School. Excursions to visit the museum became a regular event in school calendars.*

who would be entrusted with History (not including Prehistory) and the Historical Ethnography including numismatics, currency, postage stamps, tombstones and historical sites. The National Museum moved cautiously and tentatively as a result of the uncertainty brought on by its new role.

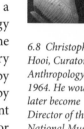
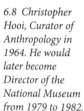

Very shortly after the National Museum's creation, the government decided to employ the museum for scientific education as well, with a focus on the technical and hard sciences. A Science and Technology Museum Committee was soon established. Then, in 1961, the government announced plans to establish a science and industry museum in a section of the National Museum to 'popularize science by holding public lectures using visual aids, films and television, thereby making more effective the acceptance of the museum as an instrument of education'.[37] Carrying the endorsement of both the Minister for Education and the Minister for Culture, it was felt such 'a museum would be of immense value to the cause of scientific and technical education for the people of Singapore' and would 'stimulate the economic life of the nation through the display of the nation's products and processes in the economic field'.[38] Eventually, the idea of situating a science and technology museum within the National Museum would grow into something bigger than the museum itself.

6.8 Christopher Hooi, Curator of Anthropology in 1964. He would later become Director of the National Museum from 1979 to 1982.

Improving the Museum … and Then Change

By the time Singapore achieved independence in 1965, the National Museum was beginning to look dated and old, leading to public calls for the government to 'modernize' it. People had started making unfavourable comparisons between the National Museum and the Muzium Negara in Kuala Lumpur which was built in 1963. On record was a statement made in July 1965 by Mrs Rosemarie Marx—a fifth-generation descendant of Raffles—that the museum was 'really in bad shape'.[39] In direct contrast to Marx's critique, Woon Wah Siang, Permanent Secretary to the Ministry of Culture, had declared just a month earlier in June 1965 that Singapore's National Museum was 'second to none in Malaysia' on account of the quality of its 'study collections' and the research generated from scholars working in them.[40] Woon further stated that a committee had been set up in 1964 to consider plans to modernise the museum's public galleries 'so that more exhibits can be effectively displayed in modern show-cases'.[41] This plan—which was part of Singapore's Second Development Plan (1966–70)—involved a major overhaul of the display cases and panels, the construction of mezzanine floors, a bridge to link the galleries, new lighting throughout and laboratory facilities.[42] By 1965, the figure had escalated to $722,600.

6.9 *Interior shot of the National Museum, 1964*

The same year, Alfred was also asked to prepare a proposal to the Ministry of Culture on how best to update and improve the National Museum to transform it into an institution with contemporary appeal. While recognising the fact that the National Museum's displays looked dated and haphazard and agreeing that an aesthetic and organisational overhaul was necessary, Alfred nevertheless emphasised the importance of its reference collection:

> Along with this change in the museum 'front' has come a change behind the scenes that, from the standpoint of collection, preservation and study is of great importance. It should be realized that the objects on exhibit represent only a small fraction of the museum's collections. It is these study collection in drawers, cabinets and on shelves (like the vast collections in the National Museum) that are the solid core of any good museum.
>
> However, any museum is unlikely to serve its functions well unless these study collections are properly maintained and adequate facilities are provided for the curators to carry out fundamental study and research.[43]

None of the plans proposed by Alfred were carried out. For unknown reasons, the Second Developmental Plan proposed in 1965 would not be finalised till December 1967 when it was announced that $470,000

would be spent on reorganising and modernising the museum so that it can 'play an effective role in hastening the "the cultural upsurge" in the Republic'.[44] In his Presidential Address of May 1968, Yusof Ishak explained the vision of the museum's future role:

6.10 Deer display and elephant skeleton, 1964

> The Library, the Museum, the Printing Office together with Radio and Television will not only inform, entertain and educate but will also stimulate patriotism, public-spiritedness and the importance of increased productivity. An appreciation for dancing, music, art and other cultural and artistic activities will be assiduously encouraged especially amongst the young.
>
> The Museum which will now come under the Ministry of Science and Technology will be modernized to serve as an educational centre displaying in vivid manner the rich political, social and economic history not only of Singapore but of the whole region of which we are the fulcrum.[45]

The Ministry of Science and Technology had been created only in April 1968,[46] and its new minister, Dr Toh Chin Chye told Parliament that when the National Museum was transferred to his ministry, plans were afoot to 'modernise the public galleries and modify the existing building to 'increase the exhibition space from 36,000 sq ft to 51,000 sq ft as well as to provide adequate accommodation for the

6.11 *Sir Solomon Hochoy, Governor-General of Trinidad &*
Tabago (top right and bottom left) visits the National Museum
in June 1964. He is being shown around by Eric Alfred, Curator
of Zoology.

6.12 Vice-President of Cambodian Council Of Ministers and Governor of Cambodian National Bank Son Sann visits the National Museum, 2 September 1966

reference collection, library and research facilities.'[47] These plans were based on the expert opinion of Dr A. N. J. Vandehoop, the Museum Advisor from UNESCO who had been consulted back in 1957, as well as the Science and Technology Museum Committee that had been established in 1961. The government also consulted Colombo Plan expert, Y. Takizawa of the National Museum of Modern Art in Tokyo, between September 1967 and June 1968, on the design and layout of exhibition cases in the public gallery.[48] However, this plan was dramatically changed towards the end of 1968 after a Gallup poll survey of museum goers. Toh explained:

> ...the Museum, as it stands at the moment, is almost a junk-shop containing a hodge-podge of exhibits. You enter one room in which you see snakes and monkeys and you enter the next room and you see old Chinese furniture and sarongs.
>
> Obviously the Museum needs a new policy and, accordingly, a Gallup poll was conducted among Museum visitors to ascertain their preferences for the various categories of exhibits in June this year. The results showed a general preference for exhibits of flora and fauna. In view of this preference and of the limitation of funds which makes it impossible to build up the Anthropological Section of the Museum, it has now been decided to develop the National Museum into a Natural History Museum.[49]

In November 1968, it was announced that experts from the United Nations Educational, Scientific and Cultural Organization (UNESCO) would help Singapore 'set up a natural science museum, a science and technology documentation centre'.[50] This was announced by Education Minister Ong Pang Boon after his return from the 15th session of UNESCO's general conference in Paris. This was followed by the government's announcement in December 1968 that the National Museum was to be transformed into a Natural History Museum at a cost of $245,000.[51]

In 1969, Alfred was again asked to prepare a plan, this time to transform the National Museum into a Natural History Museum. This lengthy proposal, which consists, in the main, plans for a physical renovation of the museum's premises, once again made the case for better research facilities and care for the Reference Collection:

> The research facilities in the Museum include the Reference collection and the Reference Library. The Reference Collection numbers some 70,000 specimens and is not only the best of its kind in South-East Asia, but is also the largest single South-East Asian

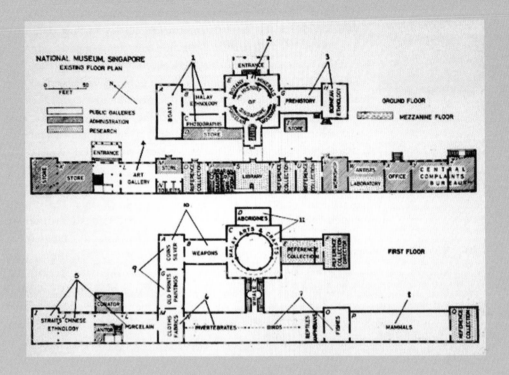

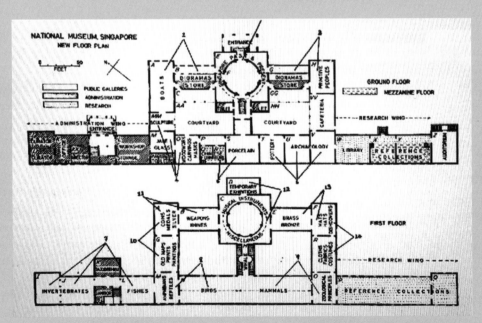

6.13 *Existing and proposed plans for the revamped National Museum, 1970*

collection. The Reference Library comprises more than 12,000 volumes and is the largest departmental library in Singapore. The Library and Reference Collection are complementary. In addition to research, the Reference Collection and the Reference Library form the basis of exhibits in the public galleries and all ancillary services. Accordingly the National Museum Board has resolved that in view of their importance the Reference Collection and Reference Library should continue to be maintained and augmented by the Museum in a South-East Asian context, to meet the needs of local Biologists.[52]

Alfred's proposal made allowance for possible changes that might be needed in light of proposals to be made by James Gardiner—a top British consultant designer—who had been engaged as UNESCO expert to conduct a feasibility survey.[53] Gardiner, who submitted his report in 1970, suggested a 'thematic treatment' in the transformation of the museum into a museum of natural history and ethnology.[54] The government, who received this report in September 1970, indicated that it was looking for a new site for the museum as well.[55]

A New Science Centre

In the meantime, a parallel development, which led to great confusion among the museum staff and the Ministry of Science and Technology officials, was taking place. In March 1970, it was announced that a 'Popular Science Centre' would be built at an estimated cost of S$5.2 million.[56] A site had yet to be selected. A month later, it was revealed that the government had set aside $1 million in the development estimates for this centre.[57]

In July 1970, an official from the Ministry of Science and Technology announced that the Science Centre would have two divisions: the first was a 'technology division to show the workings of machinery, rockets, telephones, production of paper, oil refining and the like; and the second was to be 'a life sciences division portraying man in his natural habitat at various stages of evolution, different types of animals, birds, vegetation, and even detailed models of man's body structure.'[58] The official added that, 'the National Museum with its large and diverse collections can contribute to this project' and that it was probable that 'the centre would be housed under the same roof as the museum, since this would foster greater administrative efficiency.'[59]

In September 1970, the government created the Science Centre Board by enacting the Science Centre Act.[60] The main objective behind the Science Centre was to promote science and technology through a museum/gallery site. In projecting the importance and popularity of the Science Centre, Toh estimated that the Centre would attract an average of 340,000 visitors annually, more than three times the number at the National Museum.[61] In moving the Science Centre Bill, Toh explained the rationale for subsuming a part of the National Museum under the Science Centre:

> The Government has received two studies on the modernization of the existing National Museum. In brief, the conclusion is that the Natural History Division of the National Museum should be converted to a Life Sciences Division which will display the functions of the cell; how genetics lead to the existence of species, methods of reproduction, locomotion and control of behaviour by the brain.
>
> The other proposal is that there should also be a division which will show applications of scientific principles in engineering and technology. There will be in this division a science arcade given to a series of small rooms which will cover exhibits relating to specific aspects of science, for example, the atom, sound, radar, geophysical exploration which is very relevant in Southeast Asia today. There will also be an Industrial Technology Hall dealing with the techniques made use of generally in industry.
>
> Thirdly, there will be a Communications and Transport Hall which will show the technologies used in modern communication science and transport, and will link this to the importance of Singapore as a communications and transport centre.
>
> These two divisions will constitute the Science Centre. It will be an autonomous statutory body responsible to the Ministry of Science and Technology. The Science Centre will be managed by a single body, and this body will have access to the public in the collection of funds.[62]

If the National Museum's natural history collection was to move to the new Life Sciences Division of the Science Centre, what was to become of the museum's anthropological and ethnological collection? In reply to several Members' queries on this matter in July 1970, Toh said:

> Mr Speaker, Sir, several Members have…raised questions on the future of the anthropological exhibits which are at present located in the National Museum. I would like to add that the members of the National Museum Board have recommended that the Science Centre be located at a new site. I have no doubt

that when the Science Centre Board has been appointed, it will be actively engaged in searching for a new location for the Science Centre. I am hoping that the anthropological exhibits in the National Museum will be permitted to remain there, and that they, together with paintings and other artefacts, can form an art gallery for the future. The Ministry of Culture has been responsible for raising funds for an art gallery. At present, although we are not quite certain what progress has been achieved in its fund-raising campaigns, I have no doubt that the Ministry of Culture will be only too happy to acquire the present building, which is now the National Museum, so that it can house the art gallery and whatever ancient antiques that we may collect from time to time.[63]

Speculation as to whether the Science Centre would be housed within the premises of the National Museum in Stamford Road was squashed at the end of 1972 when the government announced that a new building would be constructed in Jurong to house the Science Centre, which was to going to have two divisions: technology and life sciences.[64] All talk about transforming the National Museum into a Natural History Museum suddenly evaporated and the Science Centre was beginning to have difficulty deciding on what to do with the National Museum's natural history collection even though it was stated that its display would be 'modernized'.[65]

The idea of locating a science and technology museum *within* the National Museum was first mooted in 1961. Years later, the Science Centre Board was officially constituted on 21 November 1970 with banker Wee Cho Yaw as Chairman. Other members included Ronald Sng Ewe min (Deputy Chairman); Dr Chan Kai Lok; Dr Lawrence Chia; J. F. Conceicao; Koh Boon Piang; Dr Lam Toong Jin; Loo Ming, Sng Yew Chong and Tan Soo Yang.[66] They would draw up a blueprint for the new museum.

Shortly after this, the site for the new Centre was finalised. A 10-acre site (later increased to 16 acres) in Jurong Town Centre was selected for the Science Centre and by this time, the estimate cost for the Centre had risen considerably to S$9.5 million. Part of this budget—an estimated $300,000[67]—was expected to also cover the 'modernisation of the existing natural history exhibits in the National Museum'.[68] An architectural design competition for the new Science Centre building was announced on 22 February 1971,[69] and was won by Raymond Woo in November that year.[70] It was expected that the Centre would be complete by 1973.

The National Museum's Collections Split

On 31 March 1972, the National Museum was transferred from the Ministry of Science and Technology back to the Ministry of Culture, who would 'convert the back portion of the public galleries into an art gallery and will also run the Museum as an Anthropological and Art Museum.'[71] The museum's collection was split up with effect from April 1972 when the Ministry of Culture resumed control of the National Museum and its building in Stamford Road: it would keep its ethnological, anthropological and art collections and divest itself of its natural history collection.

The natural history exhibits in the museum, including the reference collection that was now officially referred to as the Zoological Reference Collection (ZRC), would be taken over by the Science Centre Board. Pending the completion of the new Science Centre in Jurong, the Board would occupy a portion of the museum and continue exhibiting the natural history exhibits at Stamford Road;[72] it was for a time, known as the Science Centre at the National Museum.

This collection of exhibits would 'eventually be enlarged and depicted in a modern systematic way in the new Science Centre'.[73] Shortly after its creation, the Science Centre also established a Schools Service to provide scientific programmes for teachers and students, and this too, was located at the museum's premises on Stamford Road. It later moved to Friendly Hill, off Depot Road in October 1972.[74]

Exactly what caused the radical shift in plans is not clear but may be surmised. The following facts are indisputable. First, the government, visitors and curators felt that the museum, venerable though it was, needed to be modernised and updated. Second, the government had determined that science and technology was of paramount importance, both in terms of education and in terms of its being key drivers of industrialisation. This led, among other things, to the creation of the Ministry of Science and Technology. Third, the government was prepared to invest generously in technical education and to this end, it was felt that a brand-new, ultra-modern facility—the Science Centre —would be built. The combination of these three factors had led government planners to decide to turn the National Museum into 'an Anthropological and Art Museum'[75] that promoted 'the art, culture, history and anthropology of the Southeast Asian region.'[76]

The split in the collection meant that the Zoological Reference Collection (ZRC) would have to move out of the museum's premises at Stamford Road eventually. Because a new and spacious Science Centre would be built, most people simply assumed that the entire ZRC would move there and form the core of its natural sciences exhibits. However,

(Above) 6.14 The Singapore Science Centre in Jurong. Built in 1977, it housed some of the mounted zoological specimens from the old National Museum.

6.15 Science Centre Logo

what most of the planners did not realise was the vastness and value of the Zoological Reference Collection that, for all intents and purposes, lay hidden in locked drawers, shelves and rooms. What everyone knew were the mounted zoological specimens on display, so dramatically epitomised by the skeleton of the fin whale, the gigantic seladang, G. P. Owen's crocodile and the giant leathery turtle. But these were just the proverbial tip of the iceberg.

The ZRC, as scientifically important as it was, simply could not be exhibited in its entirety and the Science Centre Board was not about to waste precious exhibition space to store the vast reference collection. Pondering this problem, former Assistant Director of the National Museum, Eric Alfred felt that a 'split' in the collection was a practical necessity:

This so-called splitting up of the collection was an idea that I thought up. I said this Science Centre was going to come up, and they've got nothing in store. So what actually was split up was not the Reference Collection. It was the mounted exhibits that people knew. I said, 'Let all of these go to the Science Centre' … So it all went! … The Science Centre was going to be only exhibition, no reference. And here was a ready, available zoological collection, which made them very happy.[77]

The Science Centre was happy to have these iconic mounted specimens for display since it was an exhibition, rather than a research, centre. These specimens—which had been put in storage—were eventually moved to the Science Centre in July 1976 in preparation for its opening in 1977.[78]

CHAPTER SEVEN

The Zoological Reference Collection
1970–77

THE DISPLAY COLLECTION was to be displayed at the new Science Centre, but what of the Zoological Reference Collection? According to Professor Bernard Tan—one of the original four members of the Science Centre Committee that had been formed in 1969[1]—the Centre never once considered taking over the Reference Collection:

> The Science Centre was initially going to be [about] merely physical science and engineering. That was the initial idea. It was Dr Toh Chin Chye's idea. He gave us parameters. That's why we called in the London Science Museum to help. They were basically a physical science and engineering and mathematics museum; a museum of engineering and industry. Somewhere along the way, the fate of the Raffles Museum got caught up in this. It was decided that there was to be another wing. I was slated to be the director of the physical science wing. Eric Alfred was going to be director of the life sciences wing, which would include the collection. The whole thing got delayed because of the incorporation of the life sciences … So, on Eric Alfred's side, the $64,000 question was: 'Do we make a modern centre or a 'museum' kind of centre? So we decided, 'No', we didn't want all that collection. We'll make it a centre that educates people on the life sciences; a 'hands-on' type of centre.

That's when the whole collection was jettisoned and in fact there
was some uncertainty as to its fate and as to who would take it.[2]

While the Science Centre took possession of the collection, it was not
prepared to absorb it in its entirety. It was certainly keen on the mounted
specimens, but not the reference collection. The ZRC's situation was
dire, and there was 'even talk about selling it or donating it.'[3] News
of this moved quickly within the international museum fraternity. On
9 July 1971, E. C. Dixon of the Foreign and Commonwealth Office's
(FCO) Overseas Development Administration wrote to Mrs Joyce
Pope of the British Museum to confirm his telephone conversation
with her about what Dr D. N. F. Hall—the FCO's Fisheries Adviser—
heard about the future of Singapore's National Museum:

> One incidental scientific item which came my way was the future
> of the National Museum. As the old Raffles Museum, the Museum
> in Singapore became world famous. Deposited in it there are very
> many type specimens from the region, particularly of birds.... I
> understand there is a move to convert the National Museum into
> a cultural museum, and to remove completely the natural history
> section with no clear idea what is to become of it, and in particular
> the type specimens. A new Raffles Museum would perpetuate a
> distinguished name.[4]

Pope discussed the matter with her colleagues at the Natural History
Museum and contacted Michael Tweedie—who had retired as
Director of the Raffles Museum back in 1957—about the likelihood of
type specimens[5] being kept in the Zoological Reference Collection as
these would be valuable to any collection. Tweedie confirmed that no
holotype specimens were 'held there'.

Even so, Arthur Percy Coleman, Secretary of the Natural History
Museum, thought it prudent to discuss the matter with Dr Ron Hedley,
his Deputy Director, who 'felt that the Keepers [Curators] might wish
to know about the possibility that the collection at the Raffles Museum
might be disposed of' and if they 'felt it would be useful, we might
approach Lord Medway, who is in Kuala Lumpur, to make appropriate
enquiries and perhaps to advise on what was available. This we would
do through Lord Cranbrook.[6] ... I would be grateful if you would let
the Director know by 20 August whether you would be interested.'[7]

This must have excited the various Keepers somewhat because
Lord Cranbrook was indeed approached. His son, Lord Medway, had
in the meantime, resigned from his position of Senior Lecturer at
the University of Malaya in Kuala Lumpur and had gone off to New

Hebrides, where he was not contactable.[8] In 1971, Dr Anthony J. Berry, of the School of Biological Sciences at the University of Malaya, spoke to Eric Alfred and 'to people in Singapore University' and determined that the 'collections in the Museum are to be handed over to the University of Singapore—officially—and the University is being given the necessary funds and other provision for handling them properly.' Berry concluded:

> My general impression is that fairly good care is being taken to house the collections suitably (perhaps even better than hitherto) and that the collections are certainly not 'going begging' or in dire need of rescue.[9]

In this instance, Berry was mistaken. Things had been in a constant state of flux and the collection was far from secure. Much more would need to be done to ensure its safety.

7.1 Lecturers and professors of Zoology Department, Bukit Timah Campus, 1969. Front row, from 2nd left: D. H. Murphy, T. E. Chua, C. F. Lim, S. H. Chuang, J. T. Harrisson (HD), D. S. Johnson, A. K. Tham, T. J. Lam, R. E. Sharma.

The Regional Marine Biological Centre

The initial alert about the collection's fate came from Hall in July 1971. Hall would have been alive to the fact of the collection as he was much involved with the key 'fish' people in the region, including Associate Professor Tham Ah Kow. Indeed, unknown to most people, it was Tham who brought the University of Singapore into the equation.

Tham Ah Kow (1913–87)[10] was born in Singapore. He was a science graduate of Raffles College, which only granted diplomas at the time. He joined the Fisheries Department upon graduation and eventually obtained a BSc by external examination from the University of London and a PhD from the University of Sydney, where he studied

with the well-known William John Dakin, Challis Professor of Zoology. In 1954, he was named Fisheries Officer and was acknowledged as Singapore's leading fisheries expert. He later became Director of the Primary Production Department in 1959. He retired from this post in 1961 aged 48 on grounds of poor health. In 1962, he established the Fisheries Biology Unit at the Department of Zoology at the University of Singapore and became its first Director. In 1971, he was promoted to the rank of Associate Professor.[11]

7.2 Common mangrove species, Episesarma chentongense, named by Raoul Serène and Soh Cheng Lam, his student, in 1967

From 1968 to 1978, under an agreement with UNESCO and the Singapore Government, a Regional Marine Biological Centre (RMBC) was established at the department, with Tham as its Director.[12] Tham would serve as the Centre's Director till his retirement in 1973 and then would be succeeded by Chuang Shou Hwa. It was in the context of Tham's work at the RMBC that he became engaged in the subject of reference collections. In 1969, there began a discussion on the feasibility of establishing a Fish Taxonomy Unit in Singapore and whether a reference collection should be created by RMBC by separating the fish collection from the National Museum's Zoological Reference Collection. Dr Raoul Serène, the UNESCO Marine Science Regional Expert for Southeast Asia,[13] stated that while it was 'not the main aim' of RMBC to establish a reference collection, its 'activities will encounter a tremendous amount of obstacles in its absence.'[14] In the course of these discussions, Tham informed Serène that Singapore's natural history collection would be split into two units:

...one unit in charge of the exhibition of Natural History with perhaps emphasis given to the educational aspect; training of teachers, popular education, mass media for natural history, etc; one unit in charge of scientific reference collections of natural history.[15]

7.3 Leo Tan and Shirley Lim collecting seashore animals from Changi beach on a class trip (circa late 1970s)

The Zoological Reference Collection, 1970

The following is a report prepared by Professor Tham Ah Kow of the Department of Zoology of the University of Singapore in May 1970 following a series of discussions on the logic of splitting the Fish Collection from the Zoological Reference Collection. This discussion resulted from the establishing of a UNESCO Regional Reference Collection for Marine Animals with a Fish Taxonomy Unit in Singapore. This is a very important document for many of Tham's suggestions and proposals were in fact adopted in the coming decades, albeit in piecemeal fashion.

Report on the Zoological Reference Collection of the National Museum[18]

This report is submitted in compliance with the Minister's instruction to me during his discussion with (a) Professor J.E. Smith, Director of the Marine Biological Laboratory at Plymouth, (b) Dr M Steyaert of the Intergovernmental Oceanographic Commission of UNESCO, and (c) Dr M Vannucci, International Curator of the Indian Ocean Biological Centre at Cochin, India, on the proposed Liaison Bureau and the Zoological Reference Collection of the National Museum of Singapore.

I visited the National Museum to inspect the extent of the collection and had a discussion with Mr Eric Alfred, Ag Director of the National Museum. The existing floor space occupied by the collection is as follows:

i.   Fishes (1,080 sq ft + 540 sq ft) = 1,620 sq ft
ii. Reptiles, Amphibia and Curstacea = 540 sq ft
iii. Birds, with some mammals (such as squirrels and rats) and some butterflies = 2,000 sq ft
iv. Mammals, mammal skins and heads (including some skeletons, some corals, some molluscs and some echinoderms) = 5,700 sq ft

Total: 9,860 sq ft

a).   The rooms in the National Museum have very high ceilings but it would be safe to assume that the storage space of about 10,000 sq ft is based on a ceiling height of 10 feet. It is conceivable that the Fish Reference Collection will be expanded in the future because the operations of the South East Asian Fisheries Research Project at Changi operating under the Ministry of National Development may be expected to yield many species which are not found in the existing collection. A large proportion of the existing fish collection consists of freshwater fish. If the Zoological Reference Collection is shifted to the University, the collections involving the other phyla will also be increased from time to time. The existing storage space is extremely cramped and there is no provision for working space. With these considerations in mind, I feel that a more realistic space requirement is 15,000 sq ft with a minimum ceiling height of 10 feet.

If it is decided to shift the Zoological Reference Collection from the National Museum to some other place, I would like to suggest that the following points be taken into consideration:–

b).   The whole collection should be shifted to one place and should not be split up and dispersed because (i) this collection has been accumulated over a long period of time and it will take just as long to assemble such a collection again, (ii) this collection is well catalogued and has been the basis of more than 400 scientific papers and publications, and (iii) the curating of each group at different locations will be more costly.

There should be a full-time qualified Curator with a supporting staff or at least two technicians and two laboratory servants. This would apply even if the whole collection were shifted to the University because the lecturers in the Zoology Department have their normal teaching duties and individual research projects to occupy them fully and cannot be expected to take responsibility for parts of the whole collection. They could however act as Curating Advisers for the particular group of animals which forms their specialty. In the same way the full-time Curator can also participate in teaching involving the particular group of animals which forms his specialty.

c).   The library can be shifted with the Reference Collection. It consists of about 12,000 volumes valued by the Ag Director of National Museum at about $100,000/–. This collection of books is housed in a room with a floor space of about 16,700 sq ft and height of 20 feet. If the new floor space has a ceiling height of 10 feet, the approximate floor space required would be in the region of 32,000 sq ft. If the Zoological Collection is shifted to the University of Singapore, the library could be put in the custody of the University Librarian in the Main Library but the loan of most of these books could be limited to bona fide research workers.

d).   There is also the question of the continued publication of the Bulletin of the National Museum which costs about $2,000 per year. According to the Ag Director of the National Museum, this Bulletin is issued in exchange for about 200 journals.

e).   The Zoological Collection and library could with advantage be shifted to the University of Singapore and renamed Raffles Museum since Raffles Museum is well-known all over the world. It is normal practice to have museums as part of a University set-up, eg at Sydney University, there is the Maclay Museum named after the benefactor Mr Maclay. It is perhaps possible that donations may be received to enable Raffles Museum to be built at the new University campus, through the British Council or some other such body.

A rough estimate of the annually recurrent expenditure for the administration of this collection as one unit within the University is as follows:–

Personal Emoluments:
1.   One Curator (Lecturer scale $920 x 40-1200/1245x45-1425 at midpoint) plus $400 VA) $19,740
2.   One Clerk/Typist (Basic $220 + $66 / VA) $3,432

3. Two Technicians (Basic $220 + $66 / VA) $6,864
4. Two Laboratory Attendants (Basic $120 + $30 / VA) $3,600
5. Provident Fund Contribution $3,861
   Total: $37,497
   O.C.A.R
6. Upkeep $2,000
7. Consumable $4,000

Total $6,000
Total Estimated Annual Expenditure $43,500

Serène questioned the wisdom of splitting the collection up but maintained that as it was 'an internal matter', he as UNESCO expert was not qualified to officially comment.[16] Tham wrote to Hoe Hwee Choo of the Ministry of Science and Technology advising against taking the Fish Reference Collection out of Singapore National Collection.[17] In April 1970, at a meeting to discuss a proposed Liaison Bureau and the wisdom of separating the Fish Reference Collection from the National Collection, Tham met with several notable marine biologists and Dr Toh Chin Chye, who was concurrently Minister for Science and Technology and Vice-Chancellor of the University of Singapore. At this meeting, Toh told Tham to prepare a report on the Zoological Reference Collection. It was submitted in May (see p.140).

Moving the Collection to the University

Nothing was heard from the ministry for another year until Tham received a note from Hoe in March 1971. She informed him that although a 'decision on the fate of the Museum's Reference Collection and Reference Library' had not yet been made, 'there is a possibility that the Collection may be accommodated at the new campus' of the university in Kent Ridge. That being the case, Tham was advised to bring this to the attention of the University of Singapore Development Unit (USDU) so that floor space requirements could be factored into the new building plans.[19]

A few days later, Tham penned a note to Reginald Quahe, Deputy Vice-Chancellor of the University of Singapore, attaching Hoe's letter and giving Quahe a background to his involvement in the discussions and how he came to draft the report for the minister. Tham prefaced his note by stating that as he was not a

7.4 Official opening of the Regional Marine Biological Centre by Toh Chin Chye, 1968. From left: C. M. Yang, A. K. Tham, Toh Chin Chee and Reginald Quahe

Head of Department and that he had no standing to communicate directly with USDU. He asked Quahe:

> If you think that the University of Singapore can entertain the proposal that the National Museum Reference Collection should be transferred to the University, could you be so good as to transmit the floor space requirements to the USDU. The estimated expenditure to be incurred in keeping the National Museum Reference Collection at the University of Singapore is of the order of $50,000 per year. However it may be possible for the Government to give a grant to the University to cover this extra expenditure.[20]

Just as Quahe was digesting the implications of Tham's note, he received a directive from Hoe, dated 30 March 1971:[21]

> Dear Sir,
> National Museum Reference Collection
> Please refer to a letter RMBC/101/69 dated 9 March 1971 written

by Prof Tham Ah Kow to the Deputy Vice-Chancellor, and copied to you.

2. The Minster for Science and Technology has directed that the Reference Collections and Reference Library of the National Museum will be transferred to the custody of the University of Singapore. It is necessary, at this juncture, to acquaint the University of Singapore Development Unit of this decision so that space requirements for housing the Collections and Library will be included in plans for the new buildings at Kent Ridge. The Minister has instructed that this matter be handed over to you so that proper arrangements will be made for the take-over.

3. You may wish to liaise directly with the Director of the National Museum for any other information in connection with this matter.

The directive came as a great shock to the university authorities who scrambled to find a way to accommodate the collection. At this time, the university occupied a beautiful but small campus on Bukit Timah Road. The campus had been built in 1928 for the Raffles College and was not intended to accommodate more than 2,000 students. By 1972, the student population of the University of Singapore was close to 10,000 and there was barely enough room for classes on campus. Indeed, classes had to be held in temporary Nissen huts erected all over the grounds of the campus. It was thus no surprise that the university did not want to be encumbered by any added strain on its already limited physical space.

In the meantime, Quahe had assigned Dr Lim Chuan Fong of the Department of Zoology to attend meetings on the transfer. In September 1971, Lim reported that he had been told that the Zoological Reference Collection would have to be removed from the National Museum building by 31 March 1972, one day before the museum was to be handed over to the Ministry of Culture. When Lim told the Committee that the existing campus at Bukit Timah had very limited space and asked if the specimens could be removed when the new campus at Kent Ridge was ready, the Working Committee 'urged the University to consider some means of temporary storage and maintenance before March 1972 as the Museum cannot wait for the specimens to be removed till the new campus is ready.'[22]

Lim had obviously also transmitted this news to Tham, who wrote to Quahe the next day to suggest two alternatives: (a) build a Nissen Hut to accommodate the whole collection on a temporary basis; or (b) convert the rooftop of the existing Zoology Department to accommodate the collection. Tham further requested Quahe to appoint Professor Chuang Shou-Hwa, Acting Head of the Zoology

Department and Member of the National Museum Board, to represent the university in future meetings involving the collection. In a note to Au Yee Pun, Principal Assistant Secretary in the Ministry of Science and Technology, Quahe informed him that Tham would cease representing the university and that Chuang would take his place starting from 1 January 1972.

In the meantime, Dr Anthony Berry of the University of Malaya had travelled to Singapore to work on the ZRC at the National Museum. He had a chance to reassess the situation and tap into the local grapevine regarding the collection's fate and its transfer to the university. What he saw and heard caused him great alarm and on his return to Kuala Lumpur in April 1972, he fired off an aerogramme to Gathorne (Lord Medway), whom he knew would be worried about the collection:

I went to S'pore recently and worked a lot in the museum. It is really folding up very soon. Exhibits (old or replacements) are to be put up at the grand new Science Centre out at Jurong (close to the bird park and more industrial splendours of S'pore). The reference materials are to go to U. of S'pore where, really, nobody wants them particularly. Some staff there look forward to augmenting their teaching collections; Nan Elliot at Nanyang[23] tried to persuade Chuang at S'pore U to 'share the loot' between the Universities. There seems to have been sufficient alarm and concern from all sides, however, to make them realise at least that lots of people are watching and that they should do their best to look after the stuff. Now they have advertised for a curator to work at the University (one only). Also it is clear that no one in Singapore (University, Alfred, Govt.) will contemplate letting the material disperse (except for minor lootings that are best left unspecified).

My assessment is that (a) international alarm has awakened basically uninterested people in S'pore to the value, even world-scale obligation, involved in the materials (b) Serène in particular has shown that the crustaceans at least are a world-important collection, now in good order with catalogues spread far and wide so that virtually anyone working on crabs must use or refer to S'pore material (c) the best hope for the collections is for all interested parties to offer support (moral and material) to Chuang S. H. (Head and Prof of Zool.) in his efforts to maintain the collections as they merit; and to seek assurances that the material will continue to be available internationally as in the past and as with other museum collections. The more of this sort of pressure cum 'support', the more they will be forced to treat it as a real set of collections. Otherwise, I'm afraid that with one poor curator and many predatory teachers, the specimens will gradually disappear.

I don't know if you are still deeply concerned about this matter but perhaps you care to hear these views just in case you can urge anyone to help to fasten a sense on commitment on the Zoologist in Singapore.[24]

Medway moved quickly. On 28 Apr 1972, he wrote to Professor Chuang Shou-Hwa telling him that he was reassured that the Collection was 'to be deposited in the safe-keeping' of his Department and that he hoped Chuang would keep it 'separate from run-of-the-mill teaching materials. He added:

There are a number of people in my field in UK who are concerned about the future of these collections. I remember Eric Alfred describing how it is sometimes an uphill struggle to persuade official circles in Singapore that such 'old' stuff can be of unique and irreplaceable value to thoroughly modern science. If you think there is any practical way by which concerned colleagues can be off [sic] assistance in any battles you may be forced to fight, please let me know.[25]

There is no record of Chuang's reply to Medway, but it was probably very clear to Chuang that the international zoological world was watching his every move carefully. Even after the decision had been made to send the collection to the university, major scholars and curators continued to enquire about its status. Dr Joe T Marshall Jr of the Division of Birds at the Museum of Natural History in Washington, DC, wrote to Dr Kenneth Jackman, first Director of the Singapore Science Centre in March 1973, expressing grave concern that the bird skins were being 'eaten from underneath by dermestid beetles during the War, so that the feathers are falling out.'[26] Marshall, who had been direct contact with Medway, ended with a plea to keep the collection in good order:

There are zoologists all over the world who have studied these Singapore collections and who feel that they are irreplaceable. Anything you can do to help keep them in shape and available will be appreciated by all.[27]

A direct offer to take over the whole of the bird collection came from Dean Arthur Amadon (1912–2003), Lamont Curator of Birds and Chairman of the Ornithology Department of the American Museum of Natural History:

> I am writing to enquire whether there is any possibility that part or all of this collection might be available for purchase. This famous and unique collection of birds, known internationally as the 'Raffles Collection', contains, as you know, very valuable scientific material. We are concerned lest the collection be broken up or scattered or subjected to mishandling by students instead of preserved for science. We have here one of the two largest collections of bird skins in the world and one rich in comparative material from the East Indies. If all or part of the Raffles Museum bird collection were to come here, it would be well cared for and very helpful in research.[28]

Mrs Yang Comes on Board

According to Timothy Barnard, Chuang, who was from the University of Singapore, and Elliot, who was from Nanyang University, 'came to an agreement to house the Zoological Reference Collection (or Raffles Collection), as it quickly came to be known, at the University of Singapore.'[29] The two institutions would later merge. It is unclear exactly how this decision was to be effected since the University of Singapore campus on Bukit Timah campus was already groaning at the seams. What we do know is that sometime between September 1971 and March 1972, a decision was made by—either by the university or by the Department of Zoology—to temporarily house the collection in five Romney huts that had previously served as workshops of the British Ministry of Public Building and Works' Department of the Environment. These had recently been handed over to the government as part of the British military withdrawal from Singapore.[30]

7.5 Mrs Yang Chang Man, the indomitable Curator of the Zoological Reference Collection

The Head of Department Chuang was obviously not pleased with this added responsibility and did little to prepare the Zoology Department for the arrival of the collection. On 12 April 1972, Professor Desmond S Johnson from the department wrote to Patrick Koh, assistant to the Vice-Chancellor, informing him that the post of Curator had been advertised twice, on 10 March and 27 March. He also informed Koh that, with respect to the alterations to the Department of Environment Workshop in Ayer Rajah Road, 'no further action has been taken and no detailed plans for conversion have been made.'[31] Tham also spoke to Koh about the lack of action and left a file note stating that Chuang 'has not left any plans with regards to the renovations and adaptations to the Department of Environment Workshop.'[32]

7.6 Wooden boxes for bird/mammal skins were stacked from floor to ceiling height. About 310 such boxes or "coffins" were used for storage till 1987, with each containing 8–30 medium to bigger size skin specimens.

In the meantime, applicants for the position of Curator were being received. Among them was Mrs Yang Chang Man, a former student of Tham Ah Kow. Yang, a biology graduate from Nanyang University, had been a teacher before signing up with the University of Singapore's Zoology Department in 1966 to complete a Diploma in Fisheries. Thereafter, Yang remained at the University of Singapore to pursue an MSc degree, which she obtained in 1972 for her study of the copepods of Singapore's waters. In 1968, when Tham became the Director of the Regional Marine Biological Centre (RMBC), he persuaded Yang to join his unit and arranged for her to transfer her Public Service Commission bond to the university.

For the next four years, Yang's main job was to study, analyse, curate, and maintain about 60,000 plankton samples collected from the South China Sea.[33] She knew from Tham that the university would soon take over the collection and that they would be advertising for a Curator in the local newspapers in March 1972. Tham—who obviously thought very highly of Yang—strongly encouraged her to apply for the position; she joined the Department of Zoology as Curator while concurrently working at the RMBC.[34]

Up to this time, Yang had never seen the museum's Reference Collection and had absolutely no idea what it looked like. On her first day of work in July 1972, she reported to Chuang and was told to see Eric Alfred at the National Museum. Yang recalled:

Chuang said that Mr Alfred would tell me and show me around the specimens, [tell] which ones will be transferred from the museum to the university. And to set up, curate and, maintain the collection. I saw him. He was expecting me. He showed me around. It was an eye-opening lesson for me because this was not open to the public but only to the researchers.... I was there the whole day. He also showed me the books that would also need to be transferred.[35]

There, Yang opened numberless drawers of specimens, huge boxes of skins and skulls, and bottles of specimens preserved in formalin. The experience was overwhelming and she was stunned by both the value of the collection and the enormity of the tasks ahead. She approached Alfred for help and he recommended that she try to employ some of the collectors and junior staff at the museum to help her since they were familiar with the collection.[36] However, the only museum staff to join the university was Augustine Raphael, who had just retired as Taxidermist from the National Museum. He worked with Yang for a year before retiring. All of Yang's other staff were young and new to the job; most of their skill and expertise in curating and preserving the collection came from reference books and helpful advice proffered by kind visiting specialists.

The Move to Ayer Rajah

Yang had no experience in curatorship or in managing a century-old historical collection, but her scientific training as a marine biologist came in useful. The task was daunting but not impossible when broken up into smaller steps. The first step was to create an inventory. There were catalogues of crabs, molluscs, and fish, and some lists of holdings in the museum but these were of no use to Yang since her task was not so much an audit, but to create an inventory of the collection at the National Museum premises before it was transferred to the university. Everything had to be painstakingly taken out, examined and recorded. This she did by hand with the help of two technicians—Raphael, who was from the museum, and Lim Keng Hua—and the lists were religiously typed up by her clerk, Woo Lee Wu. It took Yang and her team over six months to do a complete inventory of over 126,000 reference specimens,[37] and by the time they were done, it was already early 1973, almost a year after the initial deadline set for removing the collection from the museum. At the end of the process, Yang and her team found that they were custodians of 126,000 specimens, of which 4,700 were type specimens (mostly paratypes).

General view of Kg. O'Carroll Scott between Kent Ridge and
Ayer Rajah Road, looking southwards.

*7.7 File
photograph of
the Kent Ridge
site in the early
1970s, where the
new National
University of
Singapore was
sited; and where
the Zoological
Reference
Collection was
eventually sited*

The next task was to move and set up the Zoological Reference
Collection (ZRC) at Ayer Rajah. As Chuang had taken no steps to
renovate the five Romney huts—which were like airplane hangers
with some abandoned machinery and huge doors—Yang was
shocked by their bare state. There were neither shelves nor fans,
the zinc roofs were leaking, there was no ceiling or furniture and
the doors were damaged. She got the Estate Office to fix them and
install ceiling fans, and personally went out to order angle-irons
for the Department Technicians to build the shelves necessary to
house the collection.

Once the shelves were ready, Yang and her team had to pack the
collection up. There was no proper packing material available and the
team scavenged whatever they could find. Her budget for the whole
move was paltry. Yang recalls:

How was I to do this [with $750], including transportation?
Finally, we had to get packing materials (hundreds and hundreds
of cartons, wooden boxes, old newspapers) from the roadside. I
don't know how I managed it … At that time my colleagues were

very cooperative and committed. Initially the transfer was made by two technicians, two attendants, with one technician from the Department to help, and one or two labourers from the Estate Office, using the Estate Office truck. When the Estate Office truck was not available, I had to hire one from outside. At that time, I hired the labourers from PSA[38] at $10 a day, and the lorry was around $50 a day.[39]

In all, the moving of about 60 lorry loads of specimens took almost a year, and did not include the books from the museum's reference collection. Yang had been instructed by Chuang to select books that would be useful for the department or collection and bring them back to the university library. She compiled a list of more than 10,000 zoology books, reprints and periodicals to be transferred to the university. These were important reference materials, with many dating back to the 1800s. After receiving the list, Chuang made no arrangements for their transfer but instead instructed Yang to get a technician to help her manually transfer the books over with baskets. She refused, arguing that this was the University Library's responsibility. Moreover, the tremendous work of sorting and setting up the zoological collection was still far from complete.

Eventually, the books were brought back to the department (instead of the Library) over the course of a few months[40] by Dennis H ('Paddy') Murphy (b. 1931), then a Senior Lecturer in the department.[41] Yang considered Murphy one of the most supportive academics in the university, someone who truly understood the value of the collection. This was especially so after Johnson, one of the ZRC's chief champions, suddenly died of cancer in August 1972.

Unfortunately, the Romney huts were terrible for the collection. The zinc roof made the huts like ovens and temperatures indoors reached 43°C in mid-afternoon and then plummeted to about 23°C at night. These huge fluctuations were deleterious to the specimens, but this was the best Yang and her small team could do. They spent most of their time maintaining and repairing the specimens, and trying to keep out the dastardly insects, mites and mould. The conditions were far from ideal but at least the collection was safe.

The ZRC spent almost five years at Ayer Rajah and was visited by a number of scientists and researchers, some of whom became good friends with Yang. The first visitor to the ZRC at Ayer Rajah was Lord Medway, who was extremely anxious to see if the collection was safe. Many of these early visitors were to prove instrumental in the next fight to save the ZRC.

Moving To Bukit Timah

In 1976, Yang was told that the Zoological Reference Collection had to be relocated—the land on which the Romney huts stood was required for the new university hospital building. At this time, Yang came in direct contact with the University's Deputy Vice-Chancellor, Reginald Quahe, who broke all protocol and called her directly, enquiring about the collection. Quahe truly appreciated the value of the Collection. As Yang recalled: 'He gave us so much hope, and at one point, offered a location on the new campus for the collection.'[42]

Ironically, at a meeting to relocate the collection, Quahe offered the University's Estate Office Workshops at Kent Ridge Campus for housing the collection. (This is the same site that would eventually be occupied by the Lee Kong Chian Museum of Natural History, the collection's permanent home in 2015.) Unfortunately, this was not meant to be as the collection lost a friend and advocate when Quahe died of a heart attack in his office on 14 Sept 1977. The offer was eventually rejected by the Estate Officer whose decision was supported by the new Deputy Vice-Chancellor, Professor Soo Cheow Seow. It would never be mentioned again.[43]

Not only was there no space at the university's new campus to temporarily house the collection at Kent Ridge, the original plan for a new building to house the ZRC at Kent Ridge, had also evaporated by 1977. The urgent need for more and more space to accommodate an anticipated spike in student numbers at the new campus simply meant that the planned space allocation for the ZRC continuously 'tumbled down the priority list.'[44] In late 1977, Yang was informed that the original plans for the building to house the ZRC at the Kent Ridge Campus had been scrapped. Apparently, Lim Chuan Fong, the Head of the Zoology Department, had informed the university's Development Unit that it was unnecessary to make provision to house the ZRC at Kent Ridge. With Lim's assurance, the Development Unit quickly pushed the ZRC off its planning radar and the collection was once again in jeopardy.

In the meantime, Yang was given a list of vacant government buildings to consider as possibilities for ZRC's new

7.8 Garage at Dalvey Road Mess used to store part of the Zoological Reference Collection

home. Together with her laboratory technicians H. P. Wang
and H. K. Lua, Yang visited and
inspected military buildings at
Sembawang and in Portsdown
Road, the PUB pump house at
Ayer Rajah; university messes
at Dalvey Road and Dalvey
Estate, and several other
places. None of the buildings
could accommodate the entire
collection. Moreover, all the
buildings were in need of repair
and renovation.

After discussions with
Lim and the Estate Office, it
was decided that the ZRC would be split up. The mammals, birds
molluscs and insects were kept at the garage, servants' quarters and
the main building of Dalvey Road Mess, a former residential quarters
for newly-arrived faculty (and a stone's throw away from Bukit
Timah campus); the fish were kept in the basement of the University
Language Centre in the old Arts Block (staff called this place 'the
dungeon' as it had only a few small windows and was very stuffy); and
the reptiles, amphibians and invertebrates were stored on the ground
floor of Manasseh Meyer Building in the space previously occupied
by Regional Marine Biological Centre. The plankton samples were
still at RMBC at that time. They were then moved to the basement in
anticipation of their transfer to Japan in 1978 when the RMBC was
to be shut down.

*7.9 Dalvey Road
Mess*

CHAPTER EIGHT

Back from the Brink
1977–87

EVEN AS YANG Chang Man was in discussions with Reginald Quahe over where the Zoological Reference Collection might next be moved to, she decided to appeal to the many friends she had made while taking care of the collection for help in securing it. She feared that with yet another move, the collection once again faced the threat of dissipation. If well-known scientists could weigh in on the importance of the collection, perhaps its chances of survival would be heightened.

One of the first scientists to write in to the university about the collection was someone Yang had not met at that time: Professor Francis John Govier Ebling (1918–92) from the Department of Zoology of the University of Sheffield. He was, at the time, External Examiner of the Department of Zoology at the University of Singapore, and had heard about the fate of the ZRC from other faculty members. Ebling dashed off a memorandum to the department:

> I understand that the University of Singapore had received the gift of the natural history specimens from the collection of the Raffles Museum. These are of irreplaceable scientific value in a rapidly changing world, and there is currently a universal interest in the conservation of museum material. I would like to express the hope that the University will do its utmost to preserve this unique historical collection, and to display it worthily.[1]

Another scientist who would prove to be Yang's staunchest ally in the fight to save the collection was Dr David Roderick Wells, from the Department of Zoology at the University of Malaya. Wells had gone to Kuala Lumpur on a Malayan Commonwealth Graduate Scholarship in 1961 and was a former PhD student[2] of Lord Medway. He became aware of and interested in the collection 'when Lord Medway started working on a series of handbooks on Malayan birds,'[3] and met Yang when she first took charge of the collection in 1972. Wells believed this collection to be of tremendous importance from a scientific point of view:

> It is a collection of the fauna of the Malay Peninsula. When I say the Malay Peninsula, it's not just Malaya and Singapore but also South Thailand, up to the bio-geographical border, which you can think of as between 10 or 11 or 12 degrees north latitude. It didn't include Burma, because Burma was a different part of the Commonwealth administration; it was under the Indian administration. For this particular area, if you put the Tring Collection and this collection side by side, they are of equal importance; and there is no bigger collection for this area anywhere else in the world. There is nothing to compare. For the whole Peninsula, nothing touches these two collections. And if they were together, it would be instantly, Number One in the world for this area.[4]

However, he knew that the only way to make oneself heard in an economy-obsessed state was to put a monetary value to the collection. Thinking back to the desperate days of 1970s, Wells recalled:

> I was down here and there was a panic. Oh my god, the government is going to hand out several of these references skins to every school and that was going to be the end of it. Yang Chang Man had taken over, as a very young girl, with this enormous task. It's her and her alone we have to thank, for having this collection at all. So she said, 'What can we do? What can we do?' So I said, 'We know they think it's worthless, so I will write to the heads of some of the world's major museums and get them to write to the Singapore government, saying that they heard that the museum was being disbanded, and would the government be interested in a bid for purchase.' So I got Dr Les Short in the American Museum of Natural History and David Snow ... from the British Museum ...
>
> I knew all these people. So I said, 'For God's sake, cook up a letter that makes it look like you want to make a bid for it. You are interested in taking it because it is of world importance.' We wanted

to make the Singapore government realise that: (a) this is recognised globally as a collection of importance; and (b) nominally, it might be worth some money. That was just my idea. Other people said, 'No, you wouldn't make an impact whatsoever, they just weren't interested.' *But*, it certainly coincided with the dropping of the scheme to disperse the whole thing.[5]

Wells' strategy was brilliant. He fired off letters to three major museums in Europe and America: the American Museum of Natural History; the Field Museum of Natural History; and of course the British Museum (Natural History) in London. To his friend Dr Lester Leroy Short, Lamont Curator of Birds at the American Museum of Natural History, Wells wrote:

Trouble in Singapore. The government plans to build a road through the present accommodation of the old Raffles Museum collections. Unless an adequate monetary (as distinct from scientific) value can be put on them, I understand from the Curator no adequate alternative set-up will be provided. I wonder therefore if you could give us a figure for the insured value of the AMNH birds, say, per 1,000 skins or other convenient unit? Everything in Singapore has its price and the fact that the collection there is virtually untapped biologically will, alone, impress no-one important.[6]

A similar letter went off to Dr Jack Fooden (b. 1927) from the Division of Mammals at the Field Museum of Natural History in Chicago, and to Ian Courtney Julian Galbraith (b. 1925), then Head of the Sub-Department of Ornithology at the British Museum (Natural History). Yang, who knew Galbraith from earlier correspondences, also wrote to him, asking him if he could give an estimate of the value of the collection for the purposes of insurance. Galbraith sent around a memorandum to his colleagues in the other departments, requesting price estimates for a collection. This dumbfounded them since they had never been faced with such a request; and in any case, the British Museum did not insure its collections since this was considered government property.[7] Reginald William Sims (1926–2012), Head of the Annelid section of the British Museum (Natural History), offered perhaps the most sensible answer:

It is of course impossible to estimate the financial value for these *Oligochaete* collections since the only yardstick would be the price for any other collection, and I have no knowledge of such collections being offered for sale.[8]

The American curators were less nonplussed by Wells' and Yang's requests. Dr Short offered the following guidelines on pricing:

> It is exceedingly difficult to place a value on items many of which are utterly irreplaceable. The collection there contains many such specimens. One might figure some scheme such as the following: a) all type specimens that are sent out (and this is not a usual practice) are insured for $500.00 to $1,000.00, and this would be an appropriate value for each of them; b) to all unique specimens of historical-archival importance, I would assign a value of perhaps $100.00 apiece; c) for data-bearing specimens from areas in which one cannot obtain replacements (i.e., the forests are gone, or the area is politically impossible) I would guess a value of perhaps $50.00 each; d) to all other specimens bearing data, a value of $10.00; and e) to specimens with little or no data $5.00 apiece. None of these estimated values are to be taken as truly indicative of worth, for one could argue that a unique specimen that could not be replaced should be valued higher, at thousands of dollars!
>
> I have given you one set of figures treating 'values'. In practical terms, collections that are *sold* today, and they rarely are, do not come up to the value just given. It is in fact standard practice to purchase specimens comprising a collection at $3.00 to $5.00 apiece.[9]

Yang continued to collect estimates from other quarters—on mammals from Lord Medway; and insects and other arthropods from the Field Museum of Natural History—and compiled a one-page table of estimated value of the collection. It came to a staggering $8,747,853.00.[10] Based on an insurance premium of 50 cents per $1,000 value, the insurance premium for this collection alone would exceed $4,000 per annum. This was the 'valuation' Yang would present her bosses.

Unfortunately, the economic and financial argument, compelling as it was, did not help secure a proper space for the ZRC on the new campus: the authorities thought that the collection had little to contribute to Singapore's development.

Holding the Fort ... Amidst Shifting Sands

The years 1977 to 1979 were a nightmare for the ZRC. The new university campus at Kent Ridge—a part of which included the Ayer Rajah compound that the collection was forced to vacate—would not be ready till 1980. In the meantime, the Bukit Timah campus was so overcrowded that classes were being run in the most appalling conditions. Where would the ZRC go?

With more than 120,000 specimens, the collection was huge. Even Professor Tham Ah Kow, who prepared the report on the collection in 1970, had underestimated how big it was. Indeed, he had estimated that a single Nissen hut would be sufficient to accommodate the collection; it filled five Romney huts—each of which was about ten times the size of a Nissen hut—when it moved from Ayer Rajah to Bukit Timah.[11] As mentioned previously, the collection was split up for storage and housed in places with varying states of disrepair. Indeed, the conditions were so dreadful that when Crown Prince Akihito (now Emperor Akihito) of Japan visited Singapore in March 1981 and asked to visit the university to look at the goby specimens—he being one of the world's leading experts on these fish—the university had to politely decline. Instead, Yang and Lam Toong Jin, the Head of Department, had to bring the specimens to the Istana for the Prince.[12]

The splitting up of the collection meant that access for curation, maintenance, research and study was severely limited, if not well-nigh impossible in some cases. Worse still, the conditions, especially at Dalvey Mess, were very humid, and the buildings were frequently infested with termites. Fortunately, the stout teak cases housing the specimens prevented them from being destroyed. In fact, only Yang and her assistant, Lua Hui Keng, were dogged and skilful enough to work in those dreadful conditions to safeguard the collection. Oftentimes, they had to deploy very strong and poisonous insecticides and fungicides to keep specimens in good condition. Despite these harrowing conditions, Yang and her staff steadfastly and stoically provided the special care required for the delicate specimens. Most importantly, they did everything in their power to continue to make them available for study by visiting researchers and scientists. What kept them going was the importance of this scientific collection, which had to be cared and preserved for the future generations. And of course, there was the hope of the collection gaining a permanent home.

In 1979, this hope shattered when the university had yet another space crisis and rumours spread that the collection would have to be thrown out by the new university term in July 1979.[13] The government had decided to merge the Chinese-medium Nanyang University with the University of Singapore to form the National University of Singapore. As a prelude to the merger, Nanyang University students joined University of Singapore students in attending courses taught in English under a Joint Campus scheme. This led to an influx of even more students onto Bukit Timah Campus. Worse still, the new campus in Kent Ridge was nowhere near completion and every department was asked to justify its use of space. Classes were given top priority and every available space at the Bukit Timah campus would be converted to pedagogical use. Roland E Sharma, then Acting Head of the Zoology Department, recalled:

> ... it was at this time that the Collection had to contend with ever-increasing demands on its allotted space, due to an increase in student numbers resulting from a Joint Campus scheme with Nanyang University. With pressing need for additional lecture and tutorial rooms it became evident that once again, the Reference Collection would have to go.[14]

8.1 Roland E Sharma, Acting Head of the Department of Zoology from 1978 to 1981

It was Sharma who obviously bore the brunt of these pressures. A pleasant, avuncular man not prone to confrontation, Sharma felt deeply about the collection but was also powerless to fight for its preservation in Singapore.[15] In desperation, he began thinking of 'arranging for its acquisition either *in toto* by a major Museum or University overseas, or its fragmentation and dispersal amongst various institutions.'[16] News spread that the insect, fish, bird, and mammal collections would be dispersed to several Malaysian universities and institutions, including the University of Malaya, the Universiti Kebangsaan Malaysia, and Universiti Sains in Penang. He even once asked Yang if she would be prepared to 'follow the collection and work in Kuala Lumpur.'[17] Sharma felt extremely conflicted by this situation since he was well aware that the disposal of the collection 'would incur the loss of an academic heritage closely associated with Singapore and held in trust for the future' and that would in fact be 'contrary to a UNESCO Recommendation for the Protection of Movable Cultural Property, which had been adopted the previous year (1978) and one to which Singapore was a signatory.'[18] Things did not look good. In March 1979, a distraught Yang wrote to Dr Short again:

8.2 Prof Lam Toong Jin, Head of Department of the Department of Zoology from 1981 to 1996

> You may wish to be informed that the University of Singapore may not keep the old Raffles Museum collection. I learnt that the transfer of the collection to Malaysia is in the process of negotiation. The plight of this collection is unknown, but a few staff members in the dept. think that would be more safe to keep this collection in well-known museums that would preserve it for science.[19]

At the same time, Yang's international contacts pressed the Singapore authorities for news of the ZRC and even offered to absorb it into their own collections. On 28 March 1979, Robert Guestier Goelet,

President of the American Museum of Natural History prepared a letter addressed to Singapore Prime Minister Lee Kuan Yew, in which he wrote:

> Your Excellency,
> It has come to our attention that the old Raffles Museum collection of bird specimens, now in the Zoology Department of the University of Singapore, cannot longer be accommodated there. That collection is of utmost scientific importance, and ought to be place where it will be maintained properly, and be available to scientists in their studies. We ask, therefore, if the authorities presently in charge of the Raffles bird collection (perhaps Singapore University) would be interested in selling that collection to the American Museum of Natural History? If sale were to be countenanced, we would have to know the number of specimens involved before we could make an offer.
>
> The American Museum of Natural History houses the largest collection of bird specimens (1,000,000) in the world, in seven floors of a building designed for bird collections. We have plenty of space and a staff of 17 who look after the collections, which include perhaps the largest Southeast Asian bird collection. Hundreds of scientists from all over the world study in our ornithological collections, and borrow specimens from us for their investigations. Thus we are in a position to accommodate, care for, and make available to scientific use the Raffles bird collection.
>
> It is our concern for the long-range preservation and scientific utilization of the important Raffles collection of birds that prompts this inquiry.
>
> Naturally we would expect to pay for packaging and shipment of the specimens, additional to their purchase.[20]

We do not know if this letter was in fact sent but it found itself in the hands of the university's authorities. Indeed, this letter may well have come to the attention of Dr Tony Tan, Vice-Chancellor of the University and Senior Minister of State for Education at the time.

Enter the Malayan Nature Society ... and Nancy Byramji

In many ways, Sharma's hands were tied and he was too low in the university hierarchy to do more.[21] However, Sharma was also the

Chairman of the Malayan Nature Society (Singapore Branch) and it was through the Nature Society that a 'flank attack' could be undertaken. Established by J. C. Moulton in 1921, the Singapore Natural History Society was defunct by 1928 and reestablished as the Malayan Nature Society in 1940. The Singapore Branch was established in 1954 and continued to be administered as the Malayan Nature Society (Singapore Branch) even after Singapore's independence.

Sharma, who had long been involved with the society, became the Chairman of the Singapore Branch in July 1977. His committee comprised some of the most important names in Singapore's scientific community, including Dr Popuri Nageswara (P. N.) Avadhani and Professor Anne Johnson, Chairman of the Nanyang University's Department of Biology, who was the society's Honorary Secretary. Anne Johnson was the widow of Desmond S. Johnson—the late Professor of Zoology at the University of Singapore—and was as passionate about the ZRC as her late husband had been.[22] The committee had long been concerned with the fate of the ZRC but could do little more than to quietly engage the government officials privately to persuade them of the value of the collection.[23] Johnson was particularly upset over Sharma's decision to dispose of the collection to the Malaysian institutions and resolved to do something about it.[24] She decided to alert a young journalist who was a member of the society—Nancy Byramji—and get her to write an article about it.[25] At 19, Byramji had joined the *Straits Times* in 1970 and was, in her first year on the job, already covering very important stories and writing about nature.[26] By 1979, she was a seasoned and respected journalist with a strong following. Johnson provided Byramji with the background to the crisis the collection was facing[27] and on 29 April 1979, the article, 'Save our heritage: Priceless Raffles Collection may end in the dustbin unless $70,000 a year is found' appeared on the front page of the *Straits Times*.[28] Byramji had sounded the alarm bells about the university's space crunch and the 1 July 1979 deadline in the most dramatic fashion:

> Mums and dads used to take their children to the Raffles Museum to stare in awe and delight at those myriads of beautiful butterflies and insects in glass cases; at the endless varieties of birds, snakes, coral, crabs, shellfish and larger animals, stuffed and preserved.
>
> Those lucky children, grown-ups today, probably do not realise that this outstanding collection of over 100,000 wildlife specimens—many of which can no longer be seen in Singapore nor anywhere else I the world in their natural state—would well end up in the dustbin after July 1 this year.

8.3 Jaws and mounted specimen of Tiger shark in the collection. The jaw on the left belonged to a shark that was caught at Pasir Panjang in 1960, with part of a human body in its stomach.

8.4 The museum's famous Black Marlin (Makaira indica) specimen as it came out of the ZRC freezer en route to the taxidermist. The 3.3 metre long, 245 kg fish was stranded on East Coast beach on 23 November 1986, and is the only record of this fish from Singapore.

The irreplaceable, priceless, century-old Raffles Collection today poses a $70,000 problem for Singapore's zoologists—which they have only two months to solve.

They want to keep the collection intact, in Singapore—or at least within the ASEAN region.

But the collection takes up about 6,000 square metres of valuable space at the University of Singapore, and this space is needed for new lecture areas and staff rooms in the new semester to cope with the expansion of joint-campus activities.

So the collection must go.[29]

Byramji's long article detailed the history of the collection and how the tensions of space and money led to its plight. In her article she quoted 'a zoologist' as saying that the collection 'is one of the best and most complete collections of tropical fauna in the world' and another senior biologist querying 'Why has no space been allocated for this obviously valuable collection at the Kent Ridge Biological Sciences complex?' Other key quotes of the 'biologist' in her article included:

The Raffles Collection is cited in scientific reports as a very Singaporean one. It is internationally important … But it seems that because they cannot get dollars and cents out of it, the authorities are reluctant to maintain the collection, which, they claim benefits overseas people more than the locals. But it is part of our heritage, our natural history. Unless it is properly looked after by specialists who know how to curate it, as it has been for a century, the specimens could rot and be lost forever.

If the Singapore authorities are narrow-minded enough not to want such a priceless record, they should at least give it away to someone who does want to keep it, or sell it to the British Museum or Smithsonian Institute in the US which would take it over if given the chance. And what about Malaysian universities?

The university does not want to be saddled with it, largely because of costs. If the costs can be itemised and cut down, maybe we can get them to re-think. Several parties in the University of Malaysia want only portions of the collection—just the butterflies, the insects or birds. But keeping the collection is important. No one has tried to get in touch with other museums yet.

The collection was held on trust for us. It is our obligation to future generations to hold it on trust for them.[30]

In 2014, Byramji confirmed that the source in this article was in fact Sharma and that the quotes had all come from him.[31]

At this point, the situation may be summed up as follows. The University of Singapore's Zoology Department was instructed to get rid of the collection by 1 July 1979 so that room could be made for students in the coming academic year. Byramji tried getting the university to justify its actions but the 'helpful' spokesman said 'she could not comment on the situation.'[32] The biologists at the Nanyang University were keen to take over the entire collection but it did not have the money; its Director-General was 'believed to have rejected the offer because Nantah's biologists cannot justify to their administrators the $70,000 extra, and the additional staffers, that will be needed to keep the collection.'[33]

Off to Nanyang University

Byramji's article had its desired effect. The first response came from a reader, N. Sivarajah, who suggested taking 'these wildlife specimens back to the National Museum', making use of volunteer curators, and raising funds from the public to keep the collection together.[34] On 16 May 1979, Straits Times journalist Koh Yan Poh quoted Sharma as saying that while a 'number of people, representatives of government authorities, have indicated interest in keeping the collection intact in Singapore', the matter was 'far from being settled' even though 'there is some hope we may not lose it and that we may still keep the collection in place where it should be, which is in Singapore.'[35]

Less than a week later, there was a breakthrough. On 22 May 1979, during a meeting with Lu Sinclair, the formidable Registrar of the University of Singapore, Sharma and Yang, Johnson offered the Nanyang Campus as a new home for ZRC.[36] No details were disclosed since both universities declined to comment further. However, 'the certainty of a home for the collection' ensured that it would 'not be broken up' and be 'preserved for posterity.'[37]

It is not clear exactly what transpired between the authorities and the universities during this time but in July 1979, it was reported that talks were underway between the University of Singapore 'and interested parties to keep the century-old Raffles Collection intact.'[38] The Sunday Times editorial of 19 August 1979, revelling at having broken the news through Byramji's article, heaved a sigh of relief but offered an unusually critical and introspective perspective of the whole episode:

> But, amidst this moment of relief, there creeps in a feeling of shame, even fear of a sort. First, let us take the shame ... shame that a valuable

part of our scientific lore was bandied about the way it was. The Raffles Collection was treated shabbily of late, and that is shameful....

Why, one must ask, did this priceless collection pass from hand to hand as if it were a leperous object that would taint the hand that held on to it for too long? Why, in the name of science, did no one seem to realise that here was something that scientists the rest of the world over would have given their right hands to get hold of and cherish with loving care? And all this time it was in Singapore, ours ... and we did not seem to care. Yes, it did cost money—$70,000 a year to be exact—to upkeep. Yes, it did take up 6,000 square metres of space. But it is part of our heritage, even if its cold scientific value was not appreciated ... and heritage is something that can never be replaced once it is lost. It is a shame that we placed ourselves in a position that saw us come so close of losing it. Which bring us to the fear that if this could happen to the Raffles Collection, could it not happen to some other facet of our heritage? Let the Raffles Collection episode be a warning; that there are things which mean more than cold cash and quick returns; that when it comes to what was once part of our lives or the lifestyle of our land of yesteryear, it is not becoming to stand up and say that it is old, that it is time for the new to take over, that looking back is not a farsighted policy. Our past did happen ... it is history, our history—and a lot of it is our heritage. It is our responsibility to see that it is preserved for the generations to come.[39]

Having hammered out an agreement between the two universities, Roland Sharma wrote:

Fortunately, following initial consultation with Nanyang University in May of this year the immediate problems of space and location have since been solved. The very welcome offer of the use of part of the Library at Nantah (even if only on a temporary rather than a permanent basis) has relieved the situation and it is hoped to move the collection within a few months after the necessary modifications and renovation have been made to house the collection. The University of Singapore will continue to finance the Zoological Reference Collection, while the curation staff under Mrs C. M. Yang will continue to be administered as a unit within the Zoology Department. With the exception of the fish collection which will be accommodated in the Zoology Department of the University of Singapore, the remainder of the material will be transferred to Nantah where there will be adequate floor space for our specimens as well as office and laboratory facilities for

8.5 *Dried specimens of various mammals taken out in preparation for the move from its premises at Nanyang University to the National University of Singapore*

8.6 *Mammal collection at ZRC when it was at Nanyang University. Most were kept in wooden boxes arranged on the floorspace due to space constraints. These boxes were discarded in the new facility at the National University of Singapore and the specimens were transferred to modern air-tight compactors.*

8.7 Old wooden cupboards where mammal skins were kept in the ZRC when it was at Nanyang University, circa 1980s. These trays of skins were moved to compactors in new facility in the National University of Singapore.

8.8 Temporary display gallery of mounted mammals in the ZRC when it was moved to the new facility in Kent Ridge campus in the National University of Singapore

the curation staff. As a bonus the future location of the Reference Collection on the top floor of the library, provides a magnificent view across Singapore serving as a constant reminder that on geographical, historical and academic grounds this island is where the former Raffles and present Zoological Reference Collection rightfully belongs.[40]

Today, the Nature Society (Singapore) lists as one of its earliest triumphs of activism the 'preservation of the priceless historic zoological reference collection now held at the Raffles Museum of Biodiversity Research (RMBR), within today's National University of Singapore.'[41] Nanyang University had offered up the rooftop space on top of its library building to house the collection. After some renovations, Yang and her team took some six months to move the collection from Bukit Timah to its new home in 1980.

Moving Once More?

Unfortunately, just six months after they had settled in, the collection had to move again. ZRC staff was shocked and saddened by the dreadful news; they were exhausted after moving the many thousands of specimens again. Moreover, the frequent moving of the century-old, brittle specimens was extremely damaging. The old Nanyang University had merged with the University of Singapore to form the National University of Singapore (NUS) with its home in Kent Ridge. The Nanyang campus was then given over to the new Nanyang Technological Institute (NTI). Its administrators were anxious to reclaim the space for use as its administrative office.

After news spread to media that ZRC would have to make the fourth move, NUS Deputy Registrar Ling Sing Wong told the press in November 1980 that NUS was thinking of moving the specimens 'because the collection is now too far for specialists and undergraduates on the Bukit Timah campus to use for research.'[42] This announcement—made without any consultation with the Department of Zoology and certainly without consultation with Yang and her team—was a particularly nasty blow to everyone. The newspaper report ended by stating that an Australian University was interested in buying part of the collection, but that 'many scientists are horrified at such a prospect of breaking up the Collection.'[43]

Sharma, who was caught in the middle of these administrative tussles for space, was flabbergasted but soldiered on. He wrote to Christopher Hooi—who had taken over as Director of the National

Museum in 1972—in response to the request of purchasing the collection by the Australian National University:[44]

> It appears that the Collection is jeopardised on the basis of accommodation (10,000 sq ft) and maintenance and that whereas I was led to believe provision would finally be made for its location at Kent Ridge, this may not now materialise. I am accordingly looking to other sources for its survival here, within the short time I have left in office and will keep you duly informed.[45]

8.9 Packing and rearranging specimens while moving to Kent Ridge

In January 1981, Yang attended a meeting called by Sinclair, Registrar of NUS, who read out the letter from the American Museum of Natural History requesting to purchase the collection and then announced imperiously, 'The university is now convinced that ZRC is an important collection and we will not dispose of it.'[46] She also announced that a permanent home in one of the buildings at Kent Ridge campus would be provided for the collection. Sinclair told the press that the university had 'more than enough room at NTI for the zoological collection with 500 acres for an intake of only 800 students in 1982.' It was also revealed that the NTI building committee had 'decided against shifting the Institute's administration department to the Nanyang Library as originally intended.'[47] Thereafter, plans were made to accommodate the entire collection in the Science Library Building (S6) at Kent Ridge.

For seven years after the announcement (1980–87) the collection—except for the fishes—was maintained at the NTI campus and continued to serve the visiting researchers from overseas and the department staff and students. More specimens were added to the collection during this period through donations and collaborative work with visiting researchers. For example, Harold K. Voris

8.10 University
vans used for
transporting
specimens (for
fragile insects, bird
eggs, etc) from
Jurong Campus to
Kent Ridge, 1987

(Field Museum of Natural History) and William B. Jefferies (Dickinson College, Pennsylvania) visited ZRC yearly for some two decades to study goose barnacles and sea snakes from Singapore waters and nearby countries.[48] The ZRC staff was kept occupied in curating, cataloguing and maintaining the collection. They discovered that some specimens, such as molluscs, amphibians, reptiles, birds, and mammals, formerly documented and deposited at the Raffles Museum could not be traced or were lost before the collection was transferred to the university.

During this period, they also spent a lot of time repairing bird and mammal specimens that had broken heads, limbs, wings and tails. Many specimens were damaged because they had been stacked, one on top of another in huge boxes (which the staff referred to as 'coffins'). It would have been far better had they been laid out in smaller flat boxes as these smaller boxes could take the weight off the specimens. Later, all bigger or brittle specimens were placed in high quality plastic bags to eliminate further damage to the century-old specimens. The hard and valuable lessons learnt in moving and maintaining the collection helped Yang and her team to design much more suitable storage cabinets at the ZRC's new premises at Kent Ridge campus.

CHAPTER NINE

A Kent Ridge Homecoming

IN 1987, the Zoological Reference Collection moved to its 'permanent home' in three floors of the S6 building, which it shared with the Science Library of the National University of Singapore. Ironically, its new home was very close to the site of the old Romney Huts that had been its temporary home from 1972 to 1978. It took more than a year to move, unpack and rearrange the entire collection in the new home.

For the first time since its creation, the ZRC would be housed in a climate-controlled environment. The facility was air-conditioned 24 hours a day and maintained at between 22ºC and 24ºC with 55% to 60% relative humidity. The new premises for the Zoological Reference Collection, as it has officially been called since 1972, was opened on 31 October 1988 by Minister for Education, Dr Tony Tan. Also present at the launch were Michael Tweedie, former Director of the Raffles Museum, Eric Alfred, former Acting Director of the National Museum, Vice-Chancellor Lim Pin, Dean of Science Bernard Tan, and representatives of the British Council.[1]

Many zoologists, who had spent so many years worrying about the fate of the ZRC, were relieved and delighted to know it had finally found a permanent home. Gerlof Fokko Mees (1926–2013), Curator of Birds at the Rijksmuseum van Natuurlijke Historie in the Netherlands, wrote, expressing his 'great satisfaction that the collection is now adequately housed', and that 'its scientific value is recognised by the

University and by the Government of Singapore.'[2] Lord Medway, now
Earl of Cranbrook, probably spoke for all of them when he wrote,
expressing his gratitude to Yang:

> It is a personal triumph for yourself that the collection has survived
> its many moves and different homes these past years. When first
> the material was turned out of the old Raffles Museum building,
> I was not alone in thinking it lost forever. Only your persistence
> and determination has saved it through this period. Your modest
> demeanour hides a will of iron! I am so glad that you have succeeded
> in your aim.[3]

Life for Yang and her team became more settled after the move; it was
a great respite from the stresses of the first decade when they were
harried from pillar to post in a bid to hold on to the collection. Yang's
headaches were not completely over but the tide had by this time
turned in her favour and the Zoology Department was much more
appreciative of the collection and supportive of Yang's efforts. During
its first decade back at Kent Ridge, the team spent most of its time
sorting, identifying, arranging and cataloguing the collection and
repairing damaged specimens and replacing most of the formalin used
in the wet collection with alcohol since formalin is a known carcinogen.
An increasing number of scientists consulted the collection, and more
loans and exchanges of specimens were processed. Staff was also

9.2 *The opening of the Zoological Reference Collection in 1988, with Dr Tony Tan as guest-of-honour*

9.3 *Mounted animals on display at the gallery of the Zoological Reference Collection*

Resurrecting a Journal: The Raffles Bulletin of Zoology

In 1988, just as the ZRC was moving to Kent Ridge, Dennis ('Paddy') Murphy, then a Senior Lecturer at the Zoology Department, decided to revive the *Bulletin of the Raffles Museum*, which had ceased publication in 1970 following the breakup of the old National Museum. As may be recalled, the Bulletin had been started in 1928 by Cecil Boden Kloss when he was Director of the Raffles and FMS Museums. It was published regularly and continuously, except during 1942–6, as the *Bulletin of the Raffles Museum* (1928–60) and then as the *Bulletin of the National Museum* (1961–70). It was firmly established as the leading natural history journal in the region.

Murphy was denied funding from the university but managed to obtain some seed funding from the Singapore Turf Club. He renamed it '*Raffles Bulletin of Zoology*' since the Raffles Museum no longer existed. However, it was to be a continuation of its predecessor, and the volume number continued from the last publication in 1970. The first issue of the new journal was published in 1988.

9.4 *Cover of* The Raffles Bulletin of Zoology

Murphy's initial vision was to publish papers on Southeast Asian natural history, or what he called 'whole-animal zoology.' However, due to various problems, the first two issues comprised monographs: the first a catalogue of primates found in the ZRC;[4] and the second, an account of the bugs of the subfamily *Aphelocheirinae* of tropical Asia.[5] Murphy then got Peter Ng (b. 1960), his graduate student, on board to help run the journal. This changed turned things around very quickly. Ng, in his naivety, helped Murphy edit and assemble the 18 papers that appeared in the 1989 volume—the first multi-paper volume since 1970.[6]

When Murphy retired from the department in 1991, Ng 'inherited' the journal and his job as a lecturer of invertebrates. By this time, the original Turf Club funds had run out and the Department of Zoology took over the funding of the journal. Lam Toong Jin, who was Head of Department, was extremely supportive and ensured that the journal did not want for any resource so the editorial team could focus on publishing a quality international scientific journal. To this end, Ng revamped the editorial process, introduced the policy of stringent peer review by international experts, and invited an impressive number of international experts to join its editorial board. This board would eventually include top scientists like Edward O. Wilson (biodiversity), Daniel Simberloff (ecology), Herbert C. Fernando (limnology), Jack E. Randall (fish), Lipke B. Holthuis (crustacea and nomenclature) and Brian Morton (malacology and marine biology).

In 1995, the Raffles Bulletin was accepted by the Institute of Scientific Information (ISI) for inclusion in Current Contents—Agriculture, Biology & Environmental Science, a top indexing journal. It was also the first Southeast Asian journal to be listed on the Web of Science (WoS), a well-respected scientific citation indexing service and remains one of the few Asian biodiversity journals in WoS. The journal continues to reinvent itself, with a new Editor-in-Chief taking the helm every five to six years to keep it fresh and relevant. At the time of writing, the Bulletin is in its 62nd volume and is led by its fifth editor since Murphy. It has gone entirely electronic, publishes over 60 papers a year, and is fully open access, with most previous issues and supplements freely available online.

participating in field work, faunal surveys and expeditions, and hence a substantial amount of new material was added. The stability allowed Yang to develop her interest in aquatic bugs; she started spending more time on research and eventually became a top expert on them. Yang eventually retired in 2004 but continued to pursue her research interests in aquatic bugs at the ZRC.

By the time the collection moved to Kent Ridge in 1987, Roland Sharma had retired as Head of Department and was succeeded by Lam Toong Jin in 1981. Lam would remain Head of Zoology for an amazing 15 years and was an ardent supporter of the ZRC. Lam joined the University of Singapore in 1969 after graduating with a PhD from the University of British Columbia. Although he was not a taxonomist, Lam very quickly appreciated the value of the ZRC— especially its strengths in showcasing the biodiversity of the region— and resolved to support Yang and her team in all ways possible. Lam also had an excellent working relationship with Professor Koh Lip Lin, the Dean of the Faculty of Science. Koh, a chemistry professor from Nanyang University, had previously been Dean of Science at Nanyang University and was also very supportive of the ZRC. Koh got on particularly well with Lam whom he saw as a young man with plenty of drive and ambition.[7]

Space Problems Again

The first crisis Lam needed to resolve was that of space. Initially, the ZRC was only allocated three floors of the six-storey Science Library building but this was insufficient. For one thing, the collection had grown. When Yang took charge of the collection in 1972, there were some 126,000 specimens in the collection. By 1989, this had grown to 170,000 specimens. The other problem was that the allocated space was bare and there were no funds to custom-make storage cabinets for the collection. Lam went around looking for money for furniture needed to house the ZRC's collection but to no avail.

He then consulted the university's building consultants, INDECO,[8] and sat down with Yang to work out a solution. If INDECO could not provide more space to accommodate the entire collection, Lam and Yang told them that they would move the collection lock, stock and barrel from Nanyang University into the space, and that all museum storage cases[9] that could not be accommodated would be displayed in the common areas, including the foyer outside the Science Library. This worried INDECO as the library staff was already extremely nervous about being located on top of the ZRC's alcohol-preserved specimens;[10] it was like sitting on a powder keg. However, it had the desired effect of alerting everyone to the ZRC's dire need of space.

9.5 Unpacking specimens from boxes and rearranging them into the trays of compactors at the Kent Ridge premises, 1987

After much discussion, the INDECO consultant offered Lam and Yang an excellent solution. If they could incorporate their shelving system—such as a compactor system running on rails—into the building itself, the cost of this would be borne by the development budget rather than from departmental funds since it would be considered as part and parcel of the building itself. Yang got to work quickly, checking out various compactor systems. Most companies offering the product were inflexible and were unprepared to configure their systems to ZRC's special needs. They were concerned that the proposed installation would be too long and too heavy,[11] and would jam the sliding mechanism. Eventually, Yang identified a Japanese company, Kongo Co Ltd, who was very keen to custom design their compactor system to fit ZRC's special needs. A total of 276 air-tight cabinets for skin collection, operated by electrical system; and 444 open-shelf units for wet collection, by mechanical operation, were installed. This was Kongo's first time constructing air-tight cabinets for a museum, and in constructing a 36-foot-long system. It took more than a year to complete the project. The system that was eventually installed became the showcase facility both for ZRC and for Kongo.[12] Alas, most of the lovely old teak skin cabinets, and the odd-sized boxes and crates that had travelled with the collection since 1972, were discarded.

Research

In addition to ensuring that the ZRC always had sufficient funds for its operations and maintenance, Lam also contributed in many other ways. Firstly, he encouraged Yang and her staff to submit proposals for the various research grants available. With these grants, they were able to hire additional staff to help with their cataloguing and publications.

Secondly, Lam made it a point to take all his visitors to the ZRC as he was extremely proud of the collection. In no time, this almost unwanted child had become the jewel of the Zoology Department. Lam also capitalised on a collaborative programme his department had with the Japan Society for the Promotion of Science (JSPS) and brought to Singapore, Dr Hiroyuki Morioka, a senior curator of the National Science Museum of Tokyo. Morioka, a renowned ornithologist, came and worked on the bird collection and published several papers based on his studies. Later, when the visiting fellowship was over, Morioka continued to keep in touch with the Department and found his own resources to visit the ZRC to work on the specimens and published the first volume of ZRC's bird catalogue.[14] Lastly, Lam actively encouraged and financially supported field collecting and expeditions, which were regularly organised by the department in collaboration with other institutes. Through these activities, numerous specimens were collected and lodged with the ZRC. As more faculty and students began consulting the collection, they also published on the material and added to the collection.

At the same time, the ZRC also received donations of entire collections of specimens from various scientists. The original Raffles Museum collection had already been augmented in 1972 with the addition of the zoological collections of the former University of Singapore and Nanyang University, and among these were: Desmond S. Johnson's freshwater fish and crustacean collections; John Leonard Harrison's mammal collection; the fish collections of Tham Ah Kow, Samuel Tay and C. C. Lindsey; M. Nadchatram's mite collection; and Dennis H Murphy's entomological collection. In 1983, the ZRC received some 2,000 Andaman Sea and the South China Sea fish specimens from the Singapore Marine Fisheries Research Department of the Southeast Asian Fisheries Development Centre (SEAFDEC). In 1997 Constantine Herbert Fernando of the University of Waterloo donated a large collection of tropical freshwater zooplankton samples.

In the intervening years, many specimens were also contributed by research students and faculty of the Department of Biological Sciences, among whom were Peter Ng, Chou Loke Meng, Dennis H. Murphy, Kelvin K. P. Lim, and C. M. Yang. Exchange programmes with scientists, research institutions and museums abroad also brought more specimens to the collection. Finally, members of the public sometimes brought specimens to the museum as well. Through the research activities of members of the Zoology Department, the ZRC and researchers from other institutions, many new species were described and numerous scientific papers were published. At the same time, most of the type specimens were deposited in the ZRC.

C. M. Yang's report on the Zoological Reference Collection's New Home at Kent Ridge (1990)[13]

A total of about 126,000 reference specimens were transferred from the National Museum to the university in 1972. The collection represents not only specimens accumulated over the whole history of Raffles Museum since 1849, but also includes collections received from other regional research institutions. It has incorporated the reference materials of the Zoology Department of the then University of Singapore (including the important freshwater fauna collections of Desmond S. Johnson, Herbert C. Fernando, Lanna Cheng and Lim Chuan Fong; mammal collection of J. L. Harrison; fish collections of C. C. Lindsey and A. K. Tham, etc). However, in order to perform the key role of basic research and provide information on the fauna of the Southeast Asian region, the ZRC will need to expand further in the future. The number of specimens in the collection at present is about 173,000. The majority of the recent specimens have been acquired as a result of research carried out by staff and students of the department. These include about 20,000 specimens of coral and other invertebrates collected by L. M. Chou and the Reef Ecology Study Team; 1,500 mite specimens (including 76 paratypes) presented by M Nadchatram; 4,000 crab specimens collected by Peter Ng and S. Harminto; 1,000 coral fishes collected by Samuel S. W. Tay; 2,000 molluscs collected by W. H. Ho and J Sigurdsson; numerous corals by Esther G. L. Koh; coral commensals by Beverly P. L. Goh; crinoids by Grace S. Y. Lim; freshwater prawns by Samuel S. C. Chong; arthropods by D. H. Murphy; and about 3,000 specimens of insects and other aquatic animals collected from the Rompin-Endau Expedition to Peninsular Malaysia in 1989. The ZRC also benefits from a large number of valuable donations from members of public, local and foreign institutes, as well as through exchanges.

The 173,000 specimens comprise at the moment of about 15,000 mammals, 31,000 birds, 7,000 reptiles and amphibians, 31,000 fishes, 21,000 molluscs, 26,000 crabs, 10,000 prawns, 15,000 insects and 17,000 other invertebrates that are representative of the animal fauna of this region. The collection is the largest and most comprehensive in the region.

The materials of the collection are mainly used by overseas scholars as well as staff and postgraduate students of the department. It is also consulted by various government departments,

9.6 *The many trays at the Dry Collection hold thousands of bird and mammal skins, dating back to the late 1800s*

9.7 *Broadbills and pittas, some of the 30,000 bird specimens in the collection, mostly from Malaysia and Singapore. This historical collection is one of the most important such collections for the study of Malayan fauna.*

schools and members of the public. Many specimens have been loaned to overseas institutions for research purposes.

Facilities
The new premises occupy the lower three floors of the building which also accommodates the Science Library at the upper three levels. The collection is stored in 24 hours air-conditioning (22–24°C) with humidity controlled at ± 60 % r.h.

Dry collection: The bird and mammal reference skins and skulls are stored in a 230 sq. metre electrically operated mobile compactor system containing 228 units of air-tight steel cabinets (1.22 or 2.44 W x 0.60 D x 2.48 m H), with adjustable wooden trays (pl. IA, B). Collections of pinned insects, molluscs, corals and echninoderms are stored in wooden cabinets in two rooms (49 and 30 sq. metres). There are two show rooms (24 and 18 sq. metres) for visitors to view some of the mounted specimens from the former National Museum collection.

Wet collection: Spirit materials of fish, amphibians, reptiles, bats and invertebrates are stored on open shelves in a 220 sq. metre mechanically operated mobile compactor system with 534 units of steel shelving (0.91 W x 0.46 D x 2.48 m H), each with seven adjustable shelves (pl. 1 C, D). The space-saving compactors provide more storage area for future expansions of the collection. It protects specimens from light, dust, mould and pests. Ample workspace and benches are provided near the storage area for visiting researchers to examine the reference material.

9.8 Wet specimens preserved in 70% ethanol: a sea star, a flying lizard, a walking catfish, head of a Hawksbill Turtle, baby Leatherback Turtle, field frog, box crab, horseshoe crab, cockroaches and scallops

Other facilities in the premises include the following:

Research/Systematics Laboratory (54 sq. metres): equipped with general laboratory facilities for visitors. Researchers also have access to the facilities in the Department of Zoology proper.

Library (33 sq. metres): Contains a small collection of reference books, periodicals and about 3,000 articles concerning materials in the ZRC. The collection of about 10,000 zoology reference books and periodicals, as well as thousands of loose reprints, originally from the National Museum is also now housed at the Science and Zoology Libraries. These are also accessible to all visitors to the ZRC.

Cold-room (10 sq. metres).

Fumigation chamber (one cubic metre): With scrubber system for fumigating specimens as and when necessary.

Workshop and preparation room (30 sq. metres): With a fume cupboard for processing of specimens and handling of hazardous chemicals.

The Zoological Reference Collection, 1996

In the 25-year period that the ZRC was in the charge of the university, its collection almost tripled in size. In 1972, the collection had 126,000 specimens, and by 1996 the number had gone up to 330,000 specimens. Of the latter, at least 10,000 are type specimens (either holotypes or paratypes). The breakdown of the collection in 1996 is as follows:

Mammals: 15,000	Birds: 31,000	Reptiles: 3,500
Amphibians: 3,300	Fishes: 65,300	Crabs: 22,000
Corals: 3,400	Insects: 120,000	Molluscs: 35,000
Other Crustaceans: 10,000	Other Invertebrates: 30,000	Zooplankton: 14,500

Very few mammal and bird specimens were collected or donated to the Museum as Singapore's accession to the Convention on International Trade in Endangered Species of Wild Fauna and Flora (CITES) in November 1986 effectively places a ban on the shooting of mammals and birds. A few specimens were donated by the public as well as by the Jurong Bird Park and the Singapore Zoological Gardens. Among the very rare specimens collected in the 1980s and 1990s were a female Banded Leaf Monkey (*Presbytis femoralis*) which had been killed by feral dogs in the Bukit Timah forest (collected by Lua Hui Kheng in October 1987); two hornbills—*Buceros bicornis* from Sentosa on 14 May 1996, and a *Buceros vigil* in January 1992; and a 245-kg Black Marlin (*Istiompax indica*) which was stranded on East Coast Beach on 23 November 1986. The help rendered by the museum during these years to scientists from all over the world as well as through publishing their papers in the *Raffles Bulletin*, also benefited the collections. The goodwill was repaid through gifts of scientific specimens from many researchers, deposition of duplicate material in the ZRC as well as many exchanges. For example, the museum's crustacean collections benefited substantially from mutually beneficial exchange programs with the Smithsonian Institution, the Muséum National d'Histoire Naturelle in Paris, and the Australian Museum.

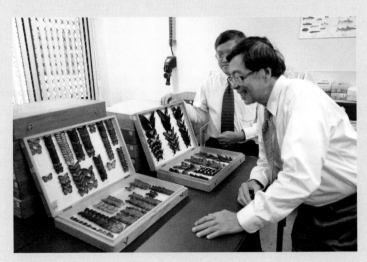

9.9 Mr Tan Teong Hean (front), with the museum's honorary butterfly curator Mr Khew Sin Khoon (back), admiring the famous W. A. Fleming Collection of Malayan Butterflies. It was purchased privately by Mr Tan and donated to the museum.

Becoming the Raffles Museum of Biodiversity Research

In 1997, Lee Soo Ying, a professor of chemistry, succeeded Bernard Tan as Dean of the Science Faculty. Lee had known about the ZRC since the time he became Vice-Dean of the Faculty in 1993, and like Lam Toong Jin, was extremely proud of it and would always take guests to visit the collection. When he took over as Dean, Lee was anxious that this national treasure should be better utilised and publicised, and proposed that the Department of Zoology consider developing a series of outreach and education programmes that would bring more people, especially school children, to the collection. It was, he said, 'like this huge box with so many wonderful things in it' and it was important that 'we open it up and share it with the public.'[15] Lee was fortunate in this enterprise in that he had the excellent support of Lam, who considered him a 'dynamic visionary'.[16]

One of the things Lee thought would help bring the ZRC a bigger audience was a public gallery. After speaking to Lam about this, Peter Ng—who was by now an Associate Professor in the Zoology Department—was tasked with preparing a paper to establish a Museum of Biodiversity Research. Ng sat down with Yang and drafted a comprehensive document entitled, 'Proposal for a Raffles Museum of Zoology: An International Research Institute for Southeast Asian Biodiversity and Establishment of a Public Display Gallery.'[17] The proposal, dated August 1997, stated that the 'primarily curatorial' mandate of the ZRC 'must be regarded as outdated' and that it was necessary to 'make museums academically stronger, more relevant for tertiary education as well as public-oriented to justify government funds.'

Ng and Yang thus proposed the establishment of a 'public exhibition area' to meet the 'increasing interest shown by the public with regards to natural history and the ZRC over the last 10 years.' Once established, the ZRC would need to be renamed as something more in keeping with its new role. Ng and Yang proposed to call it the 'Raffles Museum of Zoology' with a mission to enhance the museum's research capability, international research value, functionality in educating biology graduates as well as students in secondary and pre-tertiary education; and to increase public awareness of Singapore's and Southeast Asia's natural heritage. To do this, it was necessary to completely overhaul the museum's organisational structure, strengthen its research capability, establish a public display gallery, reorganise the facility, and appoint additional manpower to effect this plan.

9.10 Designed by its staff and volunteers, the museum's de facto logo of a civet cat on a palm leaf became synonymous with the museum

Lee was very pleased with the proposal and lent ardent support to the scheme. Initially, some members of the department were not so keen on the proposed name, arguing that going back to the name 'Raffles' smacked too much of colonial hangover. However, Ng and Lam were certain that the proposed name was timely and appropriate; after all, it was the historical name of the collection—everyone called it the Raffles Collection—and gave it instant recognition and a high profile.[18] Both Lee and Lam acted quickly. The proposal received strong support from the Vice-Chancellor, Lim Pin, and in 1998, the new facility was renamed the Raffles Museum of Biodiversity Research (RMBR).

It was now time to find someone to operationalise these proposals. As far as Lee and Lam were concerned, the only person for the job was the principal draftsman of the paper: Ng. In his letter seeking approval to establish the Raffles Museum for Biodiversity Research and the appointment of Ng as its first director, Lee Soo Ying wrote:

> The ZRC is part of the heritage of NUS, and dates back to the time of Sir Stamford Raffles. In the glory days of the 1930s and 1940s, the collection was an integral part of the Raffles Museum, one of the most famous research museums in Asia then. Today, we are ready to reassume that role and place ourselves at the forefront of biodiversity research again.[19]

As a pre-university student at Raffles Institution, Peter K. L. Ng (b. 1960) became interested in crabs. Through Roland Sharma, he had connected with Leo Tan Wee Hin (b. 1944)—who would later become his mentor—at the Department of Zoology for help in the 1978 Singapore Youth Science Fortnight competition. Much as he loved animals and the sciences, Ng did not initially apply to read science at university. After completing his GCE 'A' level examinations, he applied for a scholarship to do veterinary science, but that fell through and he then decided to read law instead. A conversation with a family friend—who was a chemistry lecturer—caused Ng to change his plans and he applied and obtained a scholarship from the Education Service to read science.[20] Ng quickly reconnected with Tan and was noticed by his many teachers, including Lam, who recalled:

> Peter [as a student] stood out because he never took any notes in class. He just sat there and stared at you. At first, I thought that he had no interest in the course, but then his marks were way up there. He graduated with a First Class Honours. I was very impressed and wanted to recruit him for the Department but found out that he was on a teaching scholarship and that he was bonded and had to

return to teach at River Valley High School, so we persuaded him to continue with his graduate studies. He wrote the biggest PhD thesis I have ever seen—two big volumes![21]

While serving his six-year bond, Ng was fortunate in that his school principal, Leong Fun Chin—who wanted him to strengthen the school's science performance—allowed him to 'turn things upside down' and gave him permission to do his PhD while teaching at the school.[22] Ng obtained his PhD in 1989[23] and joined the university in 1990.

Ng was one of the department's most active and prolific scholars. He had, as a graduate student, helped overhaul the *Raffles Bulletin of Zoology*, and was actively involved in many ZRC activities and in discovering new species. Ng would visit the Nanyang campus in Jurong almost weekly to study the crab collection as well as to deposit newly-collected specimens that he personally collected from Tuas, the Johore Straits and the East Coast.

Ng also met many visiting researchers and helped them identify crabs and other specimens and also helped in making logistical arrangements for visitors. After Ng joined the faculty at the Zoology Department, he continued to be actively involved in the activities of the ZRC and would often get his students to use the ZRC for collecting trips and in many other ways. As far as Lee and Lam was concerned, there was no better candidate for the Director's job. Yang, the ZRC's key custodian since 1972, agreed.

The Herbarium

Once the museum was established, it also took over the Department of Botany's plant collection—the Singapore University Herbarium (SINU), which had been established in 1955 by Hamish Boyd Gilliland (1911–65), Head of the Department of Botany from 1955 to 1965.[24] Gilliland, whose specialty was grasses, personally contributed 255 sheets of grass specimens to the collection. Initially, it was established as a repository of teaching and plant voucher specimens—a representative sample of a plant species used for identification—of botanical researches conducted by the faculty and students of the university.

At the time of writing, the plant collection consists of over 30,000 flowering plant specimens (including 155 wet orchid specimens); 1,660 fern specimens; 700 moss specimens; 100 liverworts specimens and 1,235 marine algal specimens. This teaching and research collection focuses mainly on the vascular and bryophyte floras of Singapore and Malaysia. Important collections include 100 moss specimens collected by the late Anne Johnson, 2,500 voucher specimens donated by Ian Mark Turner, as well as various other collections from Hsuan Keng,

Algae – Green

James Franklin Maxwell, Wee Yeow Chin, Jumali bin Kafrawi, Haji Samsuri bin Haji Ahmad, Hugh Tan Tiang Wah and Benito Ching Tan. Today, it has grown to become a large documentation of the rich plant resources of Singapore and Southeast Asia. The herbarium and its acronym, SINU, is registered with the International Association of Plant Taxonomy (IAPT) and is listed officially in the 1990 edition of the *Index Herbariorum* published by the New York Botanical Garden for IAPT.

9.11 Some of the more than 20,000 plant specimens in the museum's herbarium (SINU). The specimens are mostly from Singapore and come from the many studies and projects by staff of the university.

Institutional Reorganization and Fresh Blood

The whole-hearted support given by Lee Soo Ying and Lam Toong Jin allowed the ZRC to be organised into a proper research museum. In addition to the appointment of Peter Ng as Director, Benito Tan (b. 1948), a professor of botany and a leading expert on mosses, was appointed Deputy Director. Ng felt that Tan's appointment was important as he 'needed a botanist to add balance' to the team.[25] Tan, who excelled in teaching and outreach, remained as Deputy Director till his retirement in 2007. With additional money being allocated to the museum, Ng was able to hire some additional staff. The first two hires were N. Sivasothi and Darren Yeo, both of whom were hired to help with research, curatorial work and outreach programmes. Sivasothi—more affectionately referred to as Siva or "Otterman"— went on to found the museum volunteers group, the Toddycats in 1997. Siva and Yeo are now Senior Lecturer and Assistant Professor respectively at the Department of Biological Sciences.

Between 1988 and 1998, the ZRC had one scientific officer (C. M. Yang), three curators (Lua Hui Kheng, Kelvin Lim and Yeo Keng Loo) and one administrative officer (Greasi Simon).[26] After 1998, more curators and research officers were recruited and the

9.12 (Top left) A participant getting close to a fallen durian at a guided walk during the the 'Love Macritchie' project

9.13 (Top right) Two volunteers and graduate students, Fung Tze Kwan and Xu Weiting, conducting a Show and Tell at the National Parks' 'Festival of Biodiversity' at Vivocity in 2013

9.14 (Middle) The museum's evergreen volunteer Kok Oi Yee conducting one of the last of our public gallery tours during the museum's 'The Last Hurrah!' two weeks before it closed

9.15 (Bottom right) Designed in 2006 by the museum's chief curator, Kelvin Lim, the Toddycats mascot was derived from the common palm civet

original team was joined by Chua Keng Soon (who took care of the herbarium), Tan Swee Hee (who took over from Darren Yeo), Tan Heok Hui (who succeeded Sivasothi), Joelle Lai (who took over from Tan Swee Hee when he became Project Manager for the Lee Kong Chian Natural History Museum), Wang Luan Keng (who helmed the education unit till 2013), and Tan Siong Kiat (who succeeded the late Yeo Keng Loo). Joelle Lai took over the management of the Toddycats when Sivasothi joined the Department of Biological Sciences as a Senior Lecturer.

When Benito Tan retired in 2007, Hugh Tan, another botanist, took over as Deputy Director. Tan's forte was management and his administrative skills were much appreciated by Ng who confessed to being allergic to day-to-day operations and routine administrative tasks. Ng recalled:

> Hugh imposed administrative discipline on us all. From 2007 onwards, as we developed our research and education programs, money was flowing in, and things were happening all the time, I realised that the management structure needed balance. I was bad at day-by-day operations and routine things. I hated paperwork. Hugh on the other hand, was great at this. It was apparent to me by 2007 that we may well grow too fast and in too many directions. I have seen this happen in other organisations … Hugh put organisation back into the museum.[27]

Outreach, Education and Financial Autonomy

On 15 June 2001, Minister for Education Teo Chee Hean declared the RMBR's Public Gallery open. With that, many of the mounted specimens, which the ZRC had recovered from the Science Centre in 1985, went on display for the first time in 30 years. Given its size, only a small portion of the collection could be displayed at any one time. The gallery featured rotating exhibits curated to introduce Southeast Asian biodiversity.

In its first six months, the displays focused on ten topics: Singapore's Biodiversity in Five Kingdoms; Tropical Habitats; Surprising Singapore; The Wonderful World of Crabs; Things People Eat; Conservation Issues; Biodiversity Research by staff and students; Education & Expeditions; Raffles the Naturalist; and New Discoveries in the Region. The tiny gallery—which was mainly visited and used by faculty and NUS students—attracted an average of 400 non-University related visitors a year. To promote the gallery and its exhibits, the museum started conducting educational workshops for schools and

9.16 A bird taxidermy class conducted by the museum's education and outreach unit for schools in 2012

the public. Initially, it started on a small scale, with workshops on the fauna and flora of Singapore conducted by the museum's officers and staff like Sivasothi and Benito Tan. They also ran small classes for teachers. While these outreach programmes were extremely important and highly successful, they were not good 'business propositions.'

As Ng recalled, the staff contemplated levying a small charge for visitors but the department's senior management rejected this proposal on grounds that the university was not a corporate body; as part of the Ministry of Education, one of its obligations was to service the public. Ng was, however, able to charge the schools and the public for special, museum-type workshops though they were not profitable. Indeed, the museum lost money in every workshop it organised. They more they conducted, the more money they lost, and this weighed heavily on Ng who was convinced that this was not a sustainable model of operations.[28]

In March 2005, Ng started an experiment with two of the museum staff—Wang Luan Keng and Alvin Lok—to see if they could develop a more cost-efficient model for its educational and outreach activities. Ng explained his entrepreneurial move:

> The premise was simple—can these two people develop at least a cost-neutral educational-outreach model with schools and public. Can they create novel programs that schools and public will pay for—but at rates that cover as much of the real expenses as possible, including their salaries, expendables and so on ... So I decided to take a risk and invest in Wang and Lok—setting aside money from my reserves for their basic salaries for two years and get them to build such a model for me. If they failed, they might lose their jobs as well, and I might lose on the investment but it was a necessary experiment.[29]

Wang and Lok worked hard but the first year was disappointing. They faced a steep learning curve and for all their best efforts, the museum was still in the red. However, the team's efforts paid off in the second year when they managed to balance the books. Shortly after this, Lok left for a more secure job but Wang remained with the museum till 2013, developing a strong business model for science education that enabled the museum to recoup all costs and even allowed it to charge some overheads. He was instrumental in demonstrating the education unit's workability, and it now rejigs its programmes on a regular basis.

The museum's efforts in delivering sustainable educational and outreach programmes enabled its staff to better understand how the public viewed science and to better work with other entities to deliver meaningful and sustainable programmes. One of the boldest of such initiatives was when the Faculty of Science jointly organised the *Dinosaurs! A T-Rex Named Sue and Friends* exhibition with the Singapore Science Centre between May and August 2006. Ng recalled how the whole exhibition came about:

> In 2006, we co-organised a T-Rex exhibition with the Science Centre. The dinosaurs were from the Chicago Field Museum which we had visited and where we had good friends. They suggested that we host this exhibition, but being novices, my first inclination was to decline. But then Science Centre was game, and so was my then Dean of Science, Tan Eng Chye. In a bold move—I call it a brave move because we had no precedent—he agreed to co-sponsor the T-Rex exhibition with the Science Centre using faculty money. It took S$500,000 of faculty funds to do this. Instead of running advertisements etc., Eng Chye decided that he would invest in the exhibition instead.[30]

The exhibition was an unqualified success, but the lessons learnt proved even more valuable: working with sponsors—the Science Centre staff, especially its head of exhibitions, Clarence Sirisena—and getting the

9.17 Sir David Attenborough, with staff and students who assisted with the filming of BBC's Life in Cold Blood *in Singapore's mangroves in 2008*

finances in order. It was, for Ng, a 'game changer' and reinforced his views on museum funding.

The 'blockbuster' exhibition cost the organisers S$1 million to stage. Ordinarily, educational exhibitions were money-losing ventures, and this was expected, given the overriding educational mission. This exhibition made over S$1.2 million. The Science Faculty had an incredible outreach tool via the dinosaurs and the museum entertained and educated numerous students and alumni. Everyone involved benefitted; the Science Faculty not only recouped its S$500,000 investment, it even made a small profit from the venture. Needless to say, it was smiles all round. Tan Eng Chye was, throughout his deanship, extremely supportive of the museum and always made available funds for expeditions, staff and purchasing key collections as well.[31]

One of the visitors to the museum in 2005 was Ambassador-at-Large, Tommy Koh, who was then Chairman of the National Heritage Board. Koh had known about the collection but was stunned to see so much of the century-old collection intact. He wrote to Vice-Chancellor Shih Choon Fong and persuaded him that with this important and historical collection, the University should consider using it as the core for a natural history museum.[32] Koh's initiative was welcomed by the Science Faculty.

In late 2005, with funding from the university's Faculty of Science and American entrepreneur Frank Levison (who was working with the university's Development Office at the time) a study tour was organised. Five staff members of RMBR, including Peter Ng, left for America on a whirlwind tour to study the most successful natural history museums. They visited the California Academy of Sciences in San Francisco; the Biodiversity Institute and Natural History Museum at Kansas University; the Berkeley Natural History Museums; the Field Museum of Natural History in Chicago; the Smithsonian Institution; and the American Museum of Natural History in New York. The main object of this tour was to understand the financial models underpinning these museums. As Ng recalled

> Frank Levison challenged us to put up a business plan for the natural history museum. Initially, I was rather irritated because I always imagined that it should have been a simple case of the Government providing the funds and us putting up a good museum. Levison, who is a very successful businessman, shared with us his experiences on various American museum boards, and hammered home the importance of being funded by a good endowment plan.[33]

At the end of the trip, Ng concluded that all successful natural history museums in America had three things in common: (a) good corporate governance; (b) a good endowment plan; and, most unexpectedly, (c) dinosaurs. Reflecting on the situation at NUS, Ng felt that only the first of the three criteria could be met—NUS was certainly a well-governed establishment—but the Science Faculty did not have any endowment to speak of and they certainly had no dinosaurs. Given these conditions, the prospects for a new natural history museum lay outside the realm of possibility.[34]

That said, Ng got interested in the American model of endowed museums, feeling that the American Museum of Natural History had the best endowment management model of them all. Ng recalled:

> The museums visited in the US included both smaller university museums to state run ones (like the Smithsonian) and private establishments. It gave us a wide breath of options on how to make things work for us in the future if we built a new museum. Financial autonomy to some degree or another was drummed into us. It resonated well for us in view of our history. We decided against visiting European and Asian museums mainly because the bulk of these were state-run and therefore mainly state-funded …. That was how the old museum was run—we needed to think out of the box because the old model has problems. Frank encouraged us to look at how the Americans did it. He was right!

Initial Efforts: Education & Fundraising

The success of the museum's education programmes encouraged its staff to look upon these programmes as possible streams of revenue. For a start, the staff sought more sponsorships and donations to support the museum's many initiatives and activities in education and research. This drive began in earnest in 2006 when large companies— like ExxonMobil Asia Pacific Pte Ltd, HSBC, Shell, the Ikano Group, Keppel Corporation, Tee Yih Jia Food Manufacturing Pte Ltd— contributed to the museum's programmes. Organisations like the Ngee Ann Kongsi and Wildlife Reserves of Singapore as well as private individuals were also targeted and they all responded positively. Outreach programmes that came out of these sponsorships included: Project Semakau (2008); the Nature Explorer's Programme (2009); the Digital Nature Archive (DNA) project (2010); and the *Private Life* series of books; and the *Singapore Biodiversity: An Encyclopedia of the Natural Environment* (2011). Some of these funds were also used to

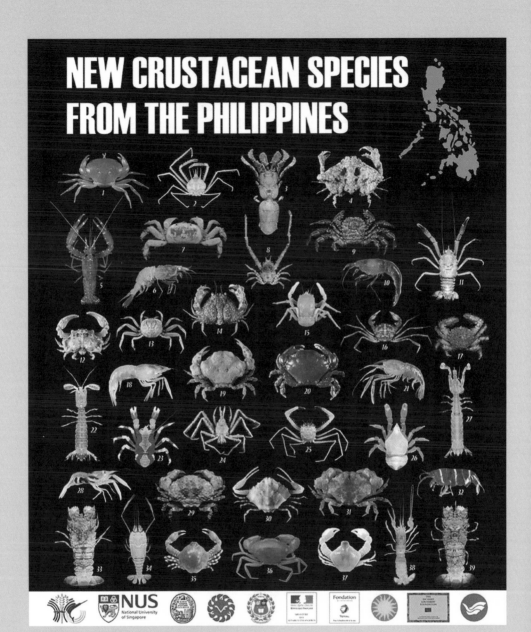

9.18 This poster was the outcome of a major expedition between 2004 and 2005 to explore the marine biodiversity of the central Philippines by the Raffles Museum, Paris Museum, Philippine National Museum, Fisheries Department of the Philippines and the National Taiwan Ocean University. Many of the species were named by Raffles Museum staff and students. Since 2004, over 100 new species have been discovered with over 100 scientific papers published.

establish an endowed research fund for Southeast Asian biodiversity research (2008) and a visiting scientist programme (*see* p.196). In all, these programmes generated several million dollars in income for the museum over the years.

The numerous education and outreach programmes put the museum in the spotlight. Many young people, who never so much as stepped barefoot on a beach, began to connect with nature in a palpable and fun way largely because of the evangelist efforts of the museum, notably by Sivasothi, in helping them learn about nature. Before long, the museum became involved in many local and even regional conservation challenges. Perhaps the most well-known of these—at least in the Singapore context—was Chek Jawa on Pulau Ubin. The museum was seen by the authorities as an honourable and level-headed advocate of environmental issues and actively engaged it in a number of national projects. For example, it has been involved with major public bodies like the National Parks Board, the Housing and Development Board and the Public Utilities Board (PUB) in undertaking environmental impact assessments, and studies into biodiversity and habitat challenges. For example, between 2006 and 2011, the PUB sponsored a five-year programme for the museum to study the biodiversity of all of Singapore's reservoirs.

Expeditions, Research, and Publications

Peter Ng had, since 1998, been anxious to restart the expeditionary and collecting work that had been at the core of the museum's activities right up till 1972. He obtained support from Lam and Lee to provide a budget for regional expeditions to collect fresh material, working with our neighbours and generating good work from the new material. Ng believed fervently in the value of expeditions:

> Expeditions are crazy things—they are stressful but they change our mindsets and pull us out of our comfort zones. They are always new and different and offer new vistas, especially working with new people. It keeps the museum researchers nimble and excited.[35]

Ng was able, through this scheme, to mount one or two small expeditions each year. Each expedition would involve between three and five local scientists, and a corresponding number of foreign counterparts as well. Each expedition would focus on tracking down selected groups of animals or habitats, and would ultimately result in good academic publications based on material collected and discoveries made. These expeditions also enabled the museum to make friends with scientists in the region and to build strong institutional ties as well.

From 1998 onwards, the RMBR, ran over 30 expeditions. Some of them were very large scale and multinational endeavours in collaboration with other universities or museums. For example, the major expeditions to the Philippines (2004–8) involved some half a dozen institutions from five countries.

Notable surveys in this period included a study of the aquatic fauna of Yunnan in southern China, the northern limit to what can be regarded as tropical Southeast Asia (2000, with the Xishuangbanna Botanical Gardens); a two-week cruise into the South China Sea through the Anambas and Natunas Islands (2002, with LIPI Indonesia); the several surveys and cruises of the Philippine Expeditions (2004–8) into the Bohol and Sulu Seas; broad-based land and sea surveys of Vanuatu (2006, with the Paris Museum & Vanuatu authorities); the Kumejima marine expedition into deep and shallow waters of the Ryukyu Islands, Japan (2009, with the University of the Ryukyus); a terrestrial and coastal survey of the famous French convict island of Con Dao in central Vietnam (2010, with the University of Hanoi); and of course, the series of successful land and sea surveys of Christmas Island and Cocos (Keeling) Islands in the Indian Ocean (2010–12, with the Australian Parks Service).

The staff of the ZRC conducted over 20 forays into Southeast Asia. Among the more significant expeditions were those to Rompin-Endau in Pahang, Malaysia (1989, with the Malayan Nature Society) to help it become a state park; North Selangor Peat Swamps in Selangor and Perak, Malaysia, to survey its unknown fauna (1991–4; with the Asian Wetland Bureau and University of Malaya); an environmental and biodiversity study of northern Pulau Bintan, Indonesia (1993-1995, under the Salim Group & LIPI, Indonesia); a park survey of Belum in Perak, Malaysia (1993, with the Malayan Nature Society); an aquatic inventory of the Lower Kinabatangan Basin in Sabah, Malaysia (1994, with the Sabah Museum); a broad based biodiversity survey of Bako National Park, Sarawak, Malaysia (1994, with the Sarawak Museum & Forestry Department of Sarawak); an aquatic study of the Maliau Basin in Sabah, Malaysia (1996, with the Sabah Museum and Universiti Sabah Malaysia); a study of freshwater animals in Brunei Darussalam (1996, with the Brunei Museum); and a study of bycatches and associated fauna in the Gulf of Thailand (1997, with Chulalongkorn University).

These expeditions resulted in the discovery of hundreds of new species of crabs, shrimps, aquatic bugs, fishes and frogs by staff and students in the ZRC and the Department of Zoology. The acquisition of new and important scientific material remains one of the strengths of the museum. The extensive material from these expeditions resulted in many other discoveries and findings as well.

9.19 Raffles Museum staff with colleagues from Universiti Malaysia Sarawak preparing to enter the forests and caves of Gunung Mulu National Park as part of a major survey in 2006

9.20 Indonesian fishermen using basket traps at Sebangau in Kalimantan. Raffles Museum staff, with researchers from the Bogor Museum, surveyed the freshwater fish fauna of this region in 2007.

All in all, over this period, the ZRC published over 300 research papers and books—a major output for such a small organisation.

When Ng took over as Director of the Museum, the ZRC had been doing well in research, having generated and published over 300 papers and books and identified hundreds of new species. Dean Lee Soo Ying charged him with further boosting the museum's research capability and output. One of Ng's initiatives was to introduce a visiting scientist programme—something he had been inspired to do after benefitting from a similar programme at the Smithsonian Institute—which allowed young scientists to spend a month or two to come to Singapore and study the collection. Ng explained:

> They come and study our collections, report on them, and in the process, add value to our material and upgrade their own research. In the process, our collections also get sorted and improved in quality. Specimens are *only* useful if sorted and studied.[36]

To this end, Ng persuaded Lam Toong Jin and Lee Soo Ying to set aside a small sum every year in his budget for a visiting scientist programme that would allow the museum to bring between two and four scientists from around the world to spend two to four weeks in Singapore to study the collection. The programme benefitted numerous scientists over the years and also allowed the museum to build a formidable regional and international network of colleagues especially since some of the beneficiaries are now senior staff or top researchers in those countries.

The *Raffles Bulletin of Zoology* was already widely-respected and well-cited and continued to gain in strength over the years. However, it did not have mass appeal, being rather academic and technical. At the same time, it did not only focus solely on Singapore but covered all of Southeast Asia. In that sense, it had far less local context and value, especially to students of Singapore's natural history. Deputy Director Hugh Tan, who was an excellent writer and editor, and who believed strongly in the need to focus on Singapore's natural history, decided to do something about this, and started an online peer-reviewed journal, *Nature in Singapore*, in 2008. Also established shortly after this was another in-house bulletin—*Singapore Biodiversity Records*—which contains short notes and observations by naturalists. It is not peer-reviewed but the museum's curators would check submissions for quality and accuracy. For larger tomes, Tan started yet another series, the *Raffles Museum Books*.

On the Map

By 2014, the research momentum that Lee Soo Ying had envisioned a decade earlier, when he asked Peter Ng to 'up the ante' in research, has been realised. Since 1998, the museum has published over 600 papers and books on all manner of animals from the material, mainly from Southeast Asia and the Indo-West Pacific. At the same time, RMBR staff and associates have discovered over a thousand new species of animals. This flurry of activities and initiatives has propelled RMBR forward as a major institution for biodiversity studies in East and Southeast Asia. Even so, more needs to be done to make the collection more accessible; Ng explained:

> Opening up our collections was important to me. Custodians of collections are very protective over specimens—as they should be. But they tend to be over-protective towards people examining the collections. I learnt from the Smithsonian that good science and good outputs require us to take risks. I was keen to loan out specimens, send out specimens all over the place, let visitors have more access to the collections. This meant taking some risks since there are some 'black sheep scientists' who can be a pain.
>
> But I was keen that we should be seen as enablers and not bureaucrats. Some of my colleagues were a bit concerned—mainly because I was going too fast, but I think it was the right decision. Of course, we made some mistakes. There were some problems and sometimes, we have even lost a few things. But the overall benefits are clear—our research profile and reputation skyrocketed!

It was becoming increasingly clear that having now secured the safety and viability of the collection, RMBR's role had to extend beyond safeguarding the collection. As a research and education outfit, more had to be done to expose it and to showcase it. The time was now ripe for a full-fledged natural history museum.

© Lee Kong Chian Natural History Museum 2015

566,500
SPECIMENS IN THE
RAFFLES MUSEUM
OF BIODIVERSITY
RESEARCH 2009

* EACH ICON REPRESENTS 1,000 SPECIMENS

SEA STARS	2,500
CORALS	5,000
REPTILES	6,800
WORMS	7,000
SPIDERS	10,000
AMPHIBIANS	12,400
MAMMALS	15,000
BIRDS	31,000
SHELLS	52,800
FISHES	80,000
CRUSTACEANS	91,000
INSECTS	108,000
PLANKTON	145,000

9.21 The museum's collection is growing continuously as a result of its research activities. Thousands of lots remain uncatalogued and will need to be sorted and studied in the future.

CHAPTER TEN

Building a World-Class
Natural History Museum

AS A MEMBER of the Museum Roundtable, the Raffles Museum of Biodiversity Research took part in International Museum Day 2009, a worldwide event to raise public awareness of the importance of museums in national development. On 24 May, the museum opened its doors, only to be greeted by a huge crowd patiently waiting to be admitted. It was estimated that some 3,000 people showed up on that day. The response was overwhelming and the staff was completely exhausted but exhilarated by the response. This set off a giddy chain of events that would eventually culminate in the decision to establish Singapore's own natural history museum.

The first salvo was fired by Jaya Kumar Narayanan, a member of the public, who wrote to the *Straits Times* Forum Page highlighting the lack of space and the inaccessibility of the museum's location.[1] Victoria Vaughan of the *Straits Times* followed up with an article highlighting the museum's predicament,[2] in which she urged Peter Ng to speak bluntly.[3] In response, Ng made an open call for a permanent natural history museum to be built, saying: 'We have an art museum, a civilisation museum, a heritage museum, but natural history is lodged in a corner of the university where no one can find it.'[4]

Ng also revealed that over the last four years (2004–8), he had been holding informal talks on the possibility of setting up a natural history museum with the Singapore Zoological Gardens, the Singapore

10.1 Museum Day 2009 visitors at the original public gallery in the Raffles Museum of Biodiversity Research

Science Centre, and the National Parks Board but 'the zoo's commercial interest and the centre's education focus was thought to be in conflict with the RMBR's research agenda, and NParks already has its work cut out looking after plant specimens.'[5]

Ten days later, on the 14 June 2009, Tan Dawn Wei wrote a very influential full page article in the *Sunday Times* openly appealing: 'Let's Have a Natural History Museum for Singapore.'[6] Ng recalled:

> [We] were inundated with visitors—thousands of people crowding into a 200 sq m gallery sited behind a small building deep in the bowels of NUS. The visitors growled—tough to find, hard to get to, gallery too small, too little displayed—complaints galore. But there was one common denominator—they all loved the place and echoed Tommy Koh's hope: Bring back Singapore's natural history museum![7]

A few months before the fateful Museum Day, Professor Leo Tan Wee Hin had returned to the university. He would eventually be a key player in the efforts to establish Singapore's natural history museum. Born on 19 October 1944 in Singapore, Tan was educated at St Joseph's Institution and then at the University of Singapore where he obtained a BSc (Hons) in Zoology. He was awarded a research scholarship and became the first local person to do a doctorate in Marine Biology. He joined the University of Singapore as a Senior Tutor in 1973 and taught at the Zoology Department from 1973 to 1986 where he rose through the ranks to become Senior Lecturer. In 1982, he was seconded to the Singapore Science Centre as its Director while concurrently

10.2 Some of the local reptiles, preserved in bottles here at the Singapore Biodiversity corner in the old public gallery of the museum, are now extinct

teaching at the university. He left the university in 1986 to become the Centre's full-time Director and remained there till 1991. During his directorship, he developed the Science Centre into one of the world's best science museums. In 1991, Tan became Foundation Dean of the School of Science at the newly-formed National Institute of Education (NIE). Three years later, he became Director of NIE and remained there till 2008 when he returned to the National University of Singapore as Professor and Director of Special Projects Dean's Office in the Faculty of Science. Recalling the sequence of events, Tan states:

> We had joined the Museum Roundtable and it was publicised. We tried twice before, but on that day, 3,000 people turned up and the papers were intrigued and wrote a report on that. And two weeks later, on the 14th of June, Tan Dawn Wei from *Straits Times* wrote a huge piece in the *Sunday Times* entitled, 'Let's have a Natural History Museum'. I read it and I said, 'Wow!' Prior to that, several people [had written in] to the Forum Page saying it [was] a good idea. 'Why is it that we don't have a natural history museum?'[8]

The Mystery Donor

The newspaper articles and the public's positive response delighted the museum but were not in themselves sufficient to establish a full-fledged natural history museum. As noted in the previous chapter, Ng had—

following his study tour of the best-run natural history museums in the world—identified three key ingredients for a natural history museum: good governance, a healthy endowment, and a dinosaur. The Raffles Museum of Biodiversity Research was certainly well-governed, but it had no endowment and it certainly had no dinosaur. Unbeknownst to the team at the museum, they were to acquire all these ingredients over the next five years. A week after Tan Dawn Wei's article was published, 'a pleasant, nondescript gentleman turned up' and spoke to Tan Swee Hee (b. 1971), a research officer at the museum, wanting to learn more about the museum. Tan brought this to the attention of his boss, Ng.

This gentleman was a rather mysterious person who only initially left a mobile contact number, and wanted to be known only as 'James.' Ng had several more meetings with James, who later said that he represented a group of anonymous donors and that they might be able to start the ball rolling in the fundraising efforts for the museum. Eventually, James told Ng that the donors were prepared to give the university $10 million to build the museum. Ng was dumbfounded but worried since everything was all so mysterious, but James assured him that there would be no doubt as to the donors' genuineness and sincerity when all final arrangements were made.

True to his word, the senior lawyer named by James as the anonymous donors' representative was a well-known alumnus and adjunct staff of the university.[9] Baffled, Ng conferred with Leo Tan— who just returned to the Science Faculty. For Tan, the donors were an important 'sign' that it was time to finally pitch for the building of a natural history museum. As a schoolboy, he had visited the old Raffles Museum many times and marvelled at its exhibits; as a young doctoral student, he spent many hours researching the Zoological Reference Collection and solemnly promised himself, 'If I could, one day, I'd like to restore the old Raffles Museum.'[10] At that point, Tan saw the donors' offer as a chance for him to help 'return the collection to the people of Singapore.'

With an offer of $10 million,[11] Tan was confident that a new museum was within striking distance. He had seen how the Dentistry Faculty building—which was mainly pre-fabricated—had been built for $17 million and surmised that that was all that was needed. Ng and Tan decided to approach the university's president, Tan Chorh Chuan, for advice on the matter.[12] 'That will do', he thought, as he went with Ng to meet the President:

> So we went to Chorh Chuan and we said, we can do it for $20 million. Assuming we have this $10 million, we need to only raise another $10 million, do we have your permission to commence fund-raising?[13]

The President's eventual reply stunned Tan and Ng. After consulting with the university's building and estates people, he told them that $10 million was not enough. They would need at least $25 to $30 million to build a respectable building. Tan and Ng would therefore need to raise another $25 million. The President added that he believed the museum to be a very worthwhile project and that he was prepared to set aside a piece of land at the soon-to-be-built University Town for it. However, he also let them know that he simply could not reserve that piece of land for them indefinitely. He gave the pair six months to raise the money.[14] This timeline was imposed as the President could not hold onto the land any longer without giving due consideration to other demands on the space by other sectors of the university.[15] They accepted the terms and staggered out of the President's office around 7 pm that evening.

Over dinner, they stared at each other and asked themselves, 'What have we just promised?' Tan recalled, 'We were just two scientists, who knew nothing about business or fund-raising, and here we were, having just committed to raising $35 million by the following June!'[16] One particularly attractive aspect of this daunting fund-raising endeavour—at least insofar as the university was concerned—was that the Singapore Government would match, dollar-for-dollar, all funds raised. This matching scheme would enable the university to build up its endowment fund that could be invested to generate core funding for the university's various institutions.

Raising the Funds

In his own inimitable and indomitable way, Tan set out to raise the remaining $25 million from various foundations, organisations and the public. It was a desperate race against time. To raise that much money so quickly, they needed as much publicity as possible. They approached Singapore Press Holdings who offered them a $5,000 donation but undertook to give the proposed museum the much-needed publicity. Tan recalls:

> It was a race against time, but the best part about it was that while we only got $5,000 in donation from Singapore Press Holdings, their reports, both *Straits Times* and *Lianhe Zaobao* … every month they gave us one story. And I said, 'Money can't buy that kind of publicity.' So I'm very grateful to SPH. They gave us the best kind of publicity we needed.[17]

By February 2010, they had only raised about $750,000. They needed a miracle. Over the next few months, they spoke to everyone they could think of. In April, Tan and Ng started their public campaign by each pledging $20,000 towards the project.[18] Their personal contribution was significant and important, for as Tan recalled, they needed 'the moral authority' to raise money from the public, and their personal commitment meant that they would not 'feel ashamed to beg'.[19] They wrote 700 letters to friends and acquaintances asking them to donate to the cause. An external fund-raising consultant told Tan that this kind of 'spray and pray' strategy was foolish and often yielded very poor results. The final response, however, was overwhelming:

> We started this in April, two months before the deadline. The public came in with almost a million dollars. We wrote over 700 letters, two-pages each. We wrote from the heart to all the people and organisations we knew. We got back 400 replies with donations … $1.00, $6.00 … we accepted any amount.' It was a simple letter. One secretary gave us half a month of her salary, and another one gave me her whole month's salary. Another technician gave $20,000 outright without even blinking. Amazing. When people put their money where their mouths are, they believe in this cause. I told Peter we had to press on. Whether we got the money or not, whether it was enough or not, we have to go ahead with it, and raised almost $1 million from the public.[20]

With $10 million from the anonymous donor and just under $1 million from the public, Tan and Ng were still $24 million short and were in a bit of a quandary as the University's Development Office had already been trying to raise funds from all the major foundations for the University Town project. Tan, whose network ranged far and wide, approached every notable he knew—ministers, businessmen, and even the President of Singapore. Another institution Ng and Tan approached was the Singapore Totalisator Board (Tote Board) but they were told that the organization did not support scientific infrastructure or museums per se. They were aware that it was a long shot, as Ng recalled:

> We had more discussions explaining that it was not just science but our national natural heritage … And we were lucky that the board's chairman at that time, Bobby Chin, was a passionate believer in natural heritage. We eventually submitted a proposal for a heritage museum to Tote Board and were delighted that it was shortlisted.[21]

After that, Ng was told that he had exactly 10 minutes to present and convince the Board to give the university the money for the museum. He prepared a very punchy and sharp set of slides and did his presentation within the allotted time—what Ng loved to call his '10 minutes for $10 million' speech—and was greeted with just a few polite questions. A few days later, the Board replied: Ng and Tan could add another $10 million to the fund.

The next $15 million came through the Lee Foundation. Tan had, at a private function, mentioned their fund-raising efforts to Dr Lee Seng Tee (b. 1923), the second son of the legendary philanthropist, Lee Kong Chian. Dr Lee invited Tan and Ng to the Lee Foundation office at the OCBC Building and asked for more details about the project. It did not take long for them to convince him of the importance of the museum and its immense heritage value for the country. Shortly after, the Lee Foundation pledged $15 million to the project. Pleased, Ng and Tan pressed on with their fundraising endeavours but were not very successful for quite some time. Some weeks before the six month deadline, both Ng and Tan received an urgent call from the Lee Foundation, saying that Dr Lee would like to meet up with them again. This got them both very worried. Why would he call them at such short notice? Ng recalled:

> 'What now?' Was it going to be bad news? We walked in there and talked to him. It was very surreal. Dr Lee looked at us and said, 'We gave you $15 million'. We said, 'Yes, thank you very much.' Then he said, 'Is $15 million enough to make it world-class? I tell you what, we'll round it up to $25 million. Make it world-class, OK?' What was there to say? It was a very short meeting and as we walked out, I looked at Leo and asked, 'Leo, what just happened?' For the first time, I actually saw Leo looking a bit stunned.'[22]

When June came around, Tan and Ng had done 'the impossible'. They had raised $46 million for the museum in the immediate aftermath of the global financial meltdown of 2008. In addition to the three big donors, who had contributed a total of $45 million toward the building of the new museum, the rest was made up of public donations.[23] With this, the university was assured of the substantial matching grant from the Singapore Government.[24] Typically, all endowment monies went into the university's central pool, but the university president made the bold move of agreeing to allow the matching funds to be tied to the museum so that it could build up an independent endowment that would take care of part of the museum's operational costs. In view of the substantial support of the Lee Foundation, it

Endowments, professorships, and scholarships

In June 2012, the Kwan Im Thong Hood Cho Temple made a very generous $2 million donation to endow a professorship for the university and its conservation programs. Rudolf Meier, the museum's Deputy Director, was named the inaugural Kwan Im Thong Hood Cho Professor for Conservation in June 2014.

On 21 March 2013, ex-president S. R. Nathan had tea with Ng and Ambassador Tommy Koh to discuss the potential of creating endowed professorships and curatorships for the upcoming museum. The meeting had been 'coincidentally arranged' such that Tan would not be present.[27] Nathan had long been a friend of the museum and had strongly supported many of its fundraising activities between 2009 and 2011. One idea that emerged was the creation of a special Professorship for Biodiversity to honour Nathan's long-time friend, Leo Tan. Nathan was excited by the idea and immediately agreed to help honour the man whom he saw as a champion of Singapore's natural environment. As this book goes to press, the Leo Tan Professorship in Biodiversity is close to realisation, thanks largely to Nathan, who single-handedly raised most of the money required.

10.3 The late Lady McNeice receiving a plaque of a crab named after her (Orcovita mcneiceae) from Ng Ngan Kee, a scholarship recipient

In October 2013, Terry McNeice, son of the late Lady Yuen-Peng McNeice, established a special Endowed Graduate Scholarship worth $2.24 million in his mother's name. This scholarship will enable the museum to help train the next generation of Singapore naturalists. Lady McNeice was a noted philanthropist and Singapore's Grand Dame of conservation and the environment. She was particularly passionate about natural history and was very keen on helping young people study biology.

was decided that the new museum would be named after its founder. Furthermore, Dr Lee Kong Chian was closely associated with the first local Chancellor of the University of Singapore (the predecessor of NUS). The new institution would now be named the Lee Kong Chian Natural History Museum. Ingredient number 2—an endowment plan—was now in place.

In May 2014, Ng—hitherto Director of RMBR—was appointed Head of the Lee Kong Chian Natural History Museum.[25] In view of the growing importance of the new museum for research and education, and its substantial endowment, NUS decided that it would be an independent academic unit in the Faculty of Science, a de facto new department. It may be recalled that Lee Kong Chian had been the philanthropist whose donation in 1953 allowed the Raffles Library to break away from the museum and be transformed into a public library. As fate would have it, through its donation to the museum, the Lee

Foundation has now helped rescue the 'other half' of the old Raffles Museum and Library from oblivion.

One evening, just as the fundraising for the museum was coming to a close, the university president urgently sought out Ng and Tan. As it was already past 6 pm and the call sounded urgent, they feared that something might have gone awry. Instead, they were met by an apologetic university president who told them that the promised site was no longer available (it was given to NUS-Yale) but that he could now offer them the site of the soon-to-be vacated Estate Office. This was music to the ears of Ng and Tan as they had concluded that this was in fact an even better site, given its greater accessibility and proximity to the University Cultural Centre, NUS Museum, and the Yong Siew Toh Conservatory of Music. Ironically, the Kent Ridge Crescent location was the exact site the late Deputy Vice-Chancellor, Reginald Quahe had offered to house the collection back in 1977.

Work on the building began on 11 January 2013, and was completed in the first quarter of 2015. The 8,500 sq m, seven-storey 'green' building was designed by renowned home-grown architect Mok Wei Wei.[26] Winner of the President's Design Award in 2007, he had worked on high-profile public projects such as the refurbishing of the National Museum as well as the Victoria Concert Hall and Theatre.

Psst … Want to Buy a Dinosaur?

One of the most important specimens from the old Raffles Museum that everyone wanted back in Singapore was the Indian Fin Whale skeleton that used to hang imposingly at the National Museum. It had been given to the Muzium Negara in Kuala Lumpur by the Science Centre in 1974 and had somehow found its way to the Labuan Marine Museum (on Labuan Island, off the coast of northern Sarawak). Getting it back would be would be extremely difficult. The museum would just have to make do with its 'treasures' and hopefully acquire something spectacular along the way.

In April 2011, fate beckoned again. Leo Tan recalls:

> … a phone call comes in from Germany: 'You want a dinosaur or not? I've got three.' We didn't know these guys, but they heard about us, through the grapevine, because very few museums were being built. Ironically, he heard about our new museum plans from a German post-doctoral student who was in Peter's laboratory at NUS at the time. We had no idea why a crab researcher was linked

to a dinosaur digger and we were surprised. It was all a coincidence ... synchronicity. But this got the conversation started.[28]

The caller represented Dinosauria International LLC, a Wyoming-based fossil company that had been excavating in Wyoming's Dana Quarry site. They had three dinosaurs for sale and they were actually 80 per cent complete. Tan was amazed; he knew that there had been very few finds of this scale and of this level of completeness over the last century. The asking price: $12 million. The offer would be good for just two months. Tan thought:

10.4 Apollonia, one of the museum's three dinosaurs, in the warehouse at the Sunbury Armory, Pennsylvania, in 2010

> At first, I was just thinking if we could even afford one, but he said, 'Think about it. If you can raise the money, buy all three. They are very good.' They were found together, whether they are a family or not, we don't know, but it would be good to have them as a group. He gave us only two months. This was in April 2011. At the end of the two months, we only had three-quarters of a million and we didn't want to ask for an extension of time because it would have committed us. We kept quiet. Lo and behold, he offered us an additional two months.[29]

The university sent a team to the dig site to check the veracity of Dinosauria's claim and see the dinosaurs in situ. Tan later visited the workplace and storehouse, paying his own way. All Dinosauria's claims proved true, and this led Tan to ask Dinosauria why they were so anxious to sell the dinosaurs to the museum since they could easily make a quicker and more profitable sale to a private collector through the auction houses. Their reply pleased him:

> They said, 'The only reason we want to sell to you is because you are the only new museum that is public and open to scientists coming to study these specimens.' Both the principals of Dinosauria are palaeontologists; and as scientists, they valued research access. That was how we started the journey. In the second two-month period; we tried every other means, and eventually we struck a deal, and we got what we needed.[30]

The three dinosaurs—diplodocid sauropods—were found together. Two of them—nicknamed Apollonia and Prince—were adults and measured 24m and 27m from head to tail, while the baby dinosaur, Twinky, measured 12m from head to tail.[31]

This was to be the beginning of their 'dinosaur campaign'. Having only recently raised $46 million for the museum, how were Tan and Ng going to raise a few more million for the dinosaurs? The Development Office was upbeat about their prospects (no one had ever tried to raise money for dinosaurs before!) but for the first few months, they did not make much headway.

In July, developer Ng Swee Hua called to enquire on the progress of the dinosaur fundraising. He had met Tan a year before, when Tan was addressing the finalists for the Singapore Environmental Achievement Award (Ng Swee Hua's Siloso Beach Resort won one of the awards). Shortly after that, Ng Swee Hua invited Tan and Ng to tour his resort and they chatted briefly about the museum fundraising. When Ng was told that little headway was being made, he pledged $500,000 to 'get the ball rolling'. This was the impetus Tan and Ng needed: they now had the necessary 'seed money' to persuade more potential sponsors to come on board.

They next approached the legal representative of the anonymous donor who gave $10 million for the museum and asked if the donor would consider giving another $1 million for the dinosaurs. The response was positive, but he had a request. Because some members of the public vociferously objected to the purchase of the dinosaurs on grounds that it had nothing to do with Singapore's natural history, the donor wanted an assurance that the purchase had 'government support'.

10.5 *The Dinosauria International (DI) team with members of the Singapore museum (LKCNHM) group at Dana Quarry, Ten Sleep, Wyoming in June 2011. From left: Dr Henry Galiano (Director, DI), Dr Sebastian Klaus (post-doctoral fellow at NUS and museum volunteer), Professor Rudolf Meier (Deputy Head, LKCNHM), Dr Tan Swee Hee (Project Manager, LKCNHM), Mr Raimund Albersdörfer (Director, DI), Ms Belinda Teo (Project Manager, LKCNHM) and Mr Derek Teo (museum volunteer).*

10.6 *The Fossilogic workshop at Orem, Utah. Professor Leo Tan examining the fossil skull of Apollonia in November 2011.*

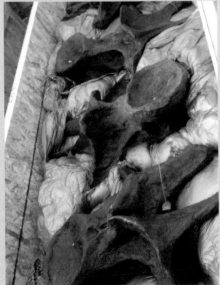

10.7 *Dinosaurs subsequently boxed up and shipped off to Singapore*

It was indeed an odd situation since dinosaurs were technically never part of Singapore's natural heritage. However, they lived over 65 million years ago, when there was no Singapore or Southeast Asia as we know it today; all the world's land masses were configured differently. In that sense then, perhaps dinosaurs are everyone's heritage!

The donors' request proved rather tricky for Ng, who decided to try his luck with Yaacob Ibrahim, the then Minister for Information, Communications and the Arts (MICA). Yaacob had been a supporter of the ZRC and the museum, and agreed to write a letter of support for the project. Another $1 million was added to the kitty.

The next big donation came from Dr Della Lee, the wife of Dr Lee Seng Gee, the eldest son of Lee Kong Chian and brother of Dr Lee Seng Tee. She was intrigued by the size and antiquity of these fossils, and after several conversations, was persuaded of the importance and value of having these dinosaurs in Singapore for everyone to see and enjoy. At Dr Della Lee's request, the managers from Dinosauria International came all the way down from the US to personally explain why the dinosaurs were valuable.

The deal was sealed. However, even with her donation, the museum was still a few million dollars short of the undisclosed agreed figure for the purchase. What else could be done to raise the remaining sum?

Once again, Tan and Ng decided to approach Dr Lee Seng Tee for advice. As with so many things that have happened over the last few years, it was a most fortuitous meeting. As Ng recalled:

> Dr Lee had back-to-back meetings with different people on the day of our meeting, and he had scheduled a discussion with Mr Philip Ng, CEO of the Far East Organisation on other matters. Our schedules overlapped. Dr Lee told Philip that he was a strong supporter of the museum and that we scientists still needed money for our dinosaurs. Turning to Leo and me, Philip asked, 'What do you guys need?' We explained that we needed a bit more money to ensure we can purchase the dinosaurs, and we also needed money to help set up a wonderful exhibit and enhance its value to education. Without a second thought he replied '$5 million should take care of all of this.' Leo and I were dumbfounded by his generosity![32]

By the end of what they called the 'dinosaur campaign', Tan, Ng, and their team had raised more than $9 million to purchase, deliver, and set up the dinosaurs and the associated displays in the gallery.[33] With this key acquisition—and not only one but three dinosaurs!—the final ingredient was now in place to guarantee the natural history museum's success.

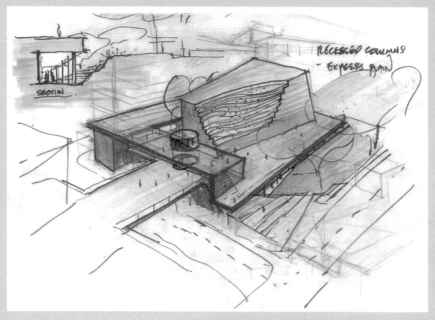

10.8 Architect's concept drawing of the new museum facility and the various access points, greenery and bridge to the Alice Lee Plaza in NUS, circa 2011

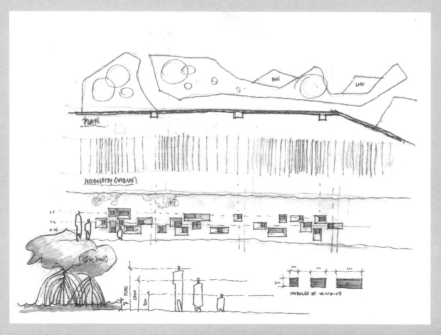

10.9 Architect's concept drawing of the mangrove garden behind the main building of the new museum, next to the teaching laboratories. This is a living exhibit of aquatic plants and can be viewed even from inside the main gallery through.

The Next Chapter ...

For Tan, Ng, and all at the RMBR, the time after 2009—or the 'psychotic years'[34]—were both inspiring and surreal. The synchronicities, coincidences and sheer luck and good fortune were just too many and too timely to be believed. Ng, a self-confessed cynic, said:

> As a scientist and cynic, I refuse to believe in the hand of fate. How can I? The scientific method frowns on destiny as an explanation for human events. But logic aside, it is very hard for a person not to harbour a nagging suspicion that the three sisters of the Moerae have intervened somewhere, somehow.[35]

Tan was much less cynical:

> I told Peter, 'I know you do not believe in divine intervention, but this cannot be purely by our own effort! There must be something ... destiny? This collection was meant for the people, so whatever you do, somehow or other, it will pan out.'[36]

The success of the dinosaur campaign did not signal the end of the museum's fund-raising exercise. And while all three 'ingredients' necessary for a successful natural history museum identified by Ng after the 2005 US study tour had been fulfilled, this was just the beginning, at least insofar as the endowment was concerned. A sizeable endowment was needed to take care of the museum's long-term finances. But as Ng remarked, 'the endgame is, and should be, for even greater financial autonomy.'[37] It was thus necessary to grow the endowment programme so that recurring expenditure in maintaining the buildings, infrastructure and collections would be well taken care of. At the same time, endowments can also help fund headcount—curators, professors and research fellows—to grow resident expertise and safeguard the interests of natural history. Endowments for researchers would also help insulate the museum from the caprice and unpredictability of administrators, politicians and funding cycles.

On 11 January 2013, the first foundation pile was driven into the ground for the construction of the Lee Kong Chian Natural History Museum. The occasion was a poignant one, witnessed by guest of honour Professor Tommy Koh, the man who started the museum on its journey in 2005, and by Dr Lee Seng Tee of the Lee Foundation, who helped make the dream a reality. Two months later, the Public Gallery at the Raffles Museum of Biodiversity Research closed its doors for

10.10 The management team of the new museum building at the ground-breaking ceremony on 11 Jan 2013. From left (upper row): *Mr Eric Too (former Associate Director, Office of Estate Development, NUS), Professor Tan Eng Chye [Deputy President (Academic Affairs) and Provost, NUS]; Professor Paul Matsudaira (Head, Department of Biological Sciences, NUS); Mr Mok Wei Wei (Managing Director, W Architects Pte Ltd); Professor Yong Kwet Yew [Vice President (Campus Infrastructure), NUS]; Mr Joseph Mullinix [Deputy President (Administration), NUS]; Professor Andrew Wee [former Dean, Faculty of Science, NUS; current Vice President (University and Global Relations), NUS]; Mr Bernard Toh (Director, Office of the President, NUS); Mr Ng Weng Pan (Director, W Architects Pte Ltd); Mr Edi Fung (Director, Development Office, NUS); Professor Leo Tan (Director, Special Projects, Faculty of Science, NUS); Dr Peter Ng (Head, Lee Kong Chian Natural History Museum, NUS) (lower row): Professor Tommy Koh (Ambassador-at Large); Mr S R Nathan (6th President, Republic of Singapore); Dr Lee Seng Tee (Director, Lee Foundation); Professor Tan Chorh Chuan (President, NUS)*

the last time, but not before being graced by a visit by President Tony Tan, to reminisce about the institution he had declared open 26 years earlier. The move from the Science Library Building to its new home will be the most massive ever. It will, hopefully, also be its last.

In 1988, when the Raffles Collection finally found a home in the ZRC in the Faculty of Science after years of nomadic existence, the man who opened it was the Education Minister Tony Tan. In April 2015, the new Lee Kong Chian Natural History Museum opened to the public—dinosaurs and all. And the man who graciously consented to launch it was again Tony Tan. Things that are meant to be have a way of coming full circle. A new chapter awaits.

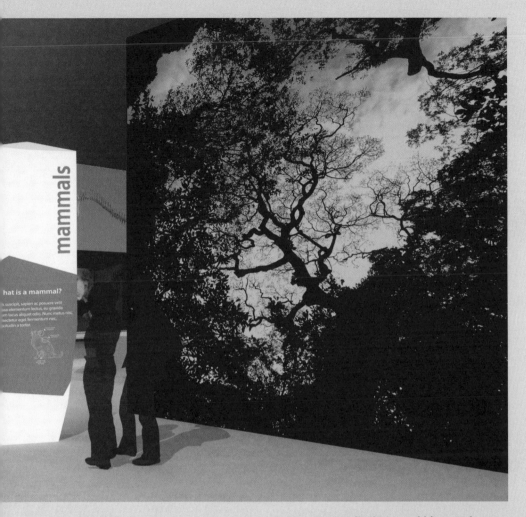

10.11 (Top) Mammals' forms and variations display presenting a variety of taxidermied specimens. Schematic design of the Ong Tiong Tat & Irene Ong Mammal Zone at the museum.

10.12 (Bottom) Large showcase showing the day and night rhythm of marine life. Final design of the Marine Cycle Zone at the museum.

Next page: Prince, lit by the moonlight, mounted outside a lab in Utah

Endnotes

Chapter 1

1. Ordinance VII of 1878.
2. Minute by Sir TS Raffles on The Establishment of a Malay College at Singapore, 1819.
3. For this account, I have relied largely on the article by Dr John Bastin, the world's leading authority on Raffles and his circle, entitled 'Raffles the Naturalist', in *The Golden Sword: Stamford Raffles and the East,* ed. Nigel Barley (London: British Museum Press, 1999) 18–29 [hereinafter 'Bastin'].
4. On the Zoological Society of London, *see* Henry Scherren, *The Zoological Society of London* (London: Cassell & Co, 1905).
5. Sophia Raffles, M*emoirs of the Life and Public Services of Sir Thomas Stamford Raffles* (London: John Murray, 1830), 4.
6. Bastin, 'Raffles the Naturalist', 19n3.
7. J. A. Bethune Cook, *Sir Thomas Stamford Raffles and Some of his Friends and Contemporaries* (London: Arthur H Stockwell, 1918), 46–7.
8. *See* John Bastin, 'William Farquhar: First Resident and Commandant of Singapore' in *Natural History Drawings: The Complete William Farquhar Collection, Malay Peninsula 1803–1818* (Singapore: EDM & National Museum of Singapore, 2010), 9–33.
9. Bastin, 'Raffles the Naturalist', 19n3.
10. Ibid., 21.
11. 'Singapore Institution', *Singapore Free Press*, 26 May 1836, 3.
12. *Singapore Free Press*, 15 Aug 1844, 2.
13. *Singapore Free Press*, 3 Oct 1844, 1.
14. *Singapore Free Press*, 20 Feb 1845, 3.
15. *Singapore Free Press*, 5 Feb 1846, 2.
16. James Richardson Logan (1819–69) was a prominent lawyer and the editor of the *Penang Gazette*. He founded the *Journal of Indian Archipelago and Eastern Asia* which he edited and published from 1847 till 1862, when it folded. He was also one of the first subscribers of the Singapore Library. He died of malaria in Penang in 1869 and is buried there.
17. 'Singapore Library', *Singapore Free Press*, 15 Feb 1849, 3.
18. Ibid.
19. Ibid.
20. Ibid.
21. Ibid.
22. *Singapore Free Press*, 5 April 1849, 2.
23. Richard Hanitsch, 'Raffles Library and Museum', in *One Hundred Years of Singapore*, vol. 1, ed. Walter Makepeace, Gilbert E Brooke & Roland St John Braddell, (London: John Murray, 1921), 536 [hereinafter 'Hanitsch'].
24. Ibid.
25. Ibid., 536–8.
26. As Hanitsch observed, there is an 'unfortunate gap in the history of the Library and Museum' as no minute book exists for the years 1866 to 1872 and this 'unluckily, coincides with a gap in the Singapore newspapers.' Ibid., 541.

27. Ibid., 541.
28. Ibid.
29. Ibid., 542n15.
30. Ibid.
31. Ibid.
32. Ibid.
33. Ibid.
34. *Straits Times,* 24 Jan 1874, 3.
35. 'Legislative Council', *Straits Times Overland Journal,* 31 Dec 1874, 8.
36. 'Legislative Council', *Straits Observer (Singapore),* 28 Dec 1874, 2.
37. 'Raffles Library and Museum' *Straits Times,* 24 Apr 1874, 1.
38. *See* 'Raffles Institution Meeting', *Straits Times Overland Journal,* 26 Mar 1874, 4.
39. *See* 'Opening of the Legislative Council', ibid., 2.
40. 'Opening of the Legislative Council', *Straits Times Overland Journal,* 26 Mar 1874, 2.
41. 'Minutes of the Legislative Council', *Straits Times,* 18 Mar 1876, 2.
42. 'Legislative Council', *Straits Times Overland Journal,* 31 Dec 1874, 8.
43. Hanitsch, 544n23.
44. 'Legislative Council', *Straits Observer,* 28 Dec 1874, 2.
45. 'Legislative Council', *Straits Times Overland Journal,* 31 Dec 1874, 8.
46. 'Municipal Commissioner', *Straits Times,* 1 Aug 1874, 2.
47. *Straits Times,* 7 Nov 1874, 2.
48. Sir Redmond Barry (1813–80) was a colonial judge in the state of Victoria. Born in Ireland, he arrived in New South Wales in April 1837 and was called to the New South Wales Bar. He was appointed Judge of the Supreme Court of Victoria in 1852. He was active in public service and was instrumental in founding the Royal Melbourne Hospital (1848); the University of Melbourne (1853); and the State Library of Australia (1854). He represented the state of Victoria at the London International Exhibition in 1862. *See* Peter Ryan, 'Barry, Sir Redmond (1813–1880)', in *Australian Dictionary of Biography* (Australia: National Centre of Biography, Australian National University, 1961).
49. *Straits Times,* 20 Feb 1875, 5.
50. Ibid.
51. Ibid.
52. 'The Philadelphia Exhibition', *Straits Times,* 27 Feb 1875, 1.
53. *See* Barbara J Black, *On Exhibit: Victorians and their Museums* (Charlottesville: University Press of Virginia, 2000).
54. Susie L. Steinbach, *Understanding the Victorians: Politics, Culture and Society in Nineteenth-Century Britain* (London: Routledge, 2012), 230.
55. 'The Philadelphia Exhibition' *Straits Times,* 27 Feb 1875.
56. 'Raffles Library and Museum', *Straits Settlements Annual Department Reports,* 1875 (Singapore: Government Printer, 1876), 2.
57. 'Ex Policeman', *Straits Observer,* 30 Jul 1875, 3.
58. *Straits Observer,* 24 Sep 1875, 2.
59. '1875', *Straits Observer,* 4 Jan 1876, 2.
60. Ibid.
61. *Straits Times,* 23 Dec 1876, 2.
62. *Straits Times,* 1 May 1875, 1.

63. 'Editorial', *Straits Times*, 30 Oct 1875, 3.
64. 'Editorial', *Straits Observer*, 4 Jan 1876, 2.
65. *Straits Times*, 18 Mar 1876, 2.
66. 'Legislative Council', *Straits Times Overland Journal*, 9 Apr 1874, 5.

Chapter 2

1. Gretchen Liu, One Hundred Years of the National Museum, Singapore 1887–1987 (Singapore: National Museum, Singapore, 1987), 21.
2. 'The Government Gazette', *Straits Times Overland Journal*, 9 Jun 1877, 14.
3. Ordinance 3 of 1877.
4. 'The Government Gazette', *Straits Times*, 23 Jun 1877, 3.
5. A very brief biographical sketch of Dennys is found in *Dictionary of British and Irish Botanists and Horticulturists*, ed. Ray Desmond (London: Taylor & Francis, and the Natural History Museum, 1994), 203.
6. *Straits Times Overland Journal*, 4 Aug 1877, 1.
7. *Straits Times*, 25 Aug 1877, 5.
8. *Straits Times Overland Journal*, 15 Sep 1877, 12.
9. *Straits Times*, 15 Sep 1877, 3.
10. *Straits Times*, 1 Sep 1877, 3.
11. *Straits Times*, 20 Oct 1877, 4.
12. 'Mrs Anson's Farewell Garden Party', *Straits Times Overland Journal*, 2 Nov 1877, 15.
13. Ibid.
14. *Straits Times*, 17 Nov 1877, 3.
15. *Straits Times Overland Journal*, 2 Nov 1877, 13.
16. *Straits Times*, 17 Nov 1877, 3.
17. Ibid.
18. Ibid.
19. 'Report on the Library and Museum for the Year 1876', *Straits Settlements Annual Department Reports 1877* (Singapore: Government Printing Office, 1878).
20 Ibid.
21 Ibid.
22. J. E. Gray, 'Presidential Address to the Botany and Zoology Section at the 34th Meeting of the British Association for the Advancement of Science', *Report of the Thirty-Fourth Meeting of the British Association for the Advancement of Science* (London: John Murray, 1865), 76–9.
23. *Straits Times Overland Journal*, 14 Mar 1878, 9.
24. *Straits Times*, 23 Mar 1878, 5.
25. *Straits Times Overland Journal*, 27 Jul 1878, 6.
26. Report on the Raffles Library and Museum for the Year 1886 (Singapore: Government Printers, 1887).
27. *Straits Times Overland Journal*, 2 Jul 1879, 7.
28. *Straits Times Overland Journal*, 21 Nov 1878, 7.
29. *Straits Times Weekly Issue*, 29 Dec 1883, 2.
30. Report on the Raffles Library and Museum for 1883 (Singapore: Government Printer, 1884).

31. *Straits Times Weekly*, 2 Apr 1883, 2.
32. Today, this crab is called *Macrocheira kaempferi*.
33. 'Report on the Raffles Library and Museum for the Year 1886', *Straits Settlements Annual Department Reports for the Year 1886* (Singapore: Government Printers, 1887), 90.
34. 'Report on the Raffles Library and Museum for 1879', *Straits Settlements Annual Department Reports for the Year 1879* (Singapore: Government Printer, 1880), 2.
35. Ibid.
36. *See* generally Bonnie Tinsley, Gardens of Perpetual Summer: The Singapore Botanic Gardens (Singapore: National Parks Board & Singapore Botanic Gardens, 2009), 31–2.
37. Bonnie Tinsley, ibid., 32.
38. *Straits Times*, 23 Nov 1878, 2.
39. Ordinance VII of 1878.
40. *Straits Times Overland Journal*, 19 May 1878, 2.
41. *Straits Times*, 25 May 1878, 4.
42. *Straits Times*, 6 Jul 1878, 1.
43. 'Report on the Raffles Library and Museum for the Year 1878', *Straits Settlements Annual Department Reports for the Year 1878* (Singapore: Government Printer, 1879), 3.
44. *Straits Times Overland Journal*, 7 Feb 1879, 2.
45. Ibid.
46. 'Comment on the Report on the Raffles Library and Museum for the Year 1878', *Straits Times Overland Journal*, 2.
47. 'From the Daily Times, 10 May', *Straits Times Overland Journal*, 15 May 1880, 5.
48. 'Report on the Raffles Library and Museum for 1880', *Straits Settlements Annual Department Reports for the Year 1880* (Singapore: Government Printer, 1881).
49. 'Announcements', *Straits Times,* 20 Mar 1883, 3.
50. 'Boys' School', *Straits Times Weekly Issue*, 20 Jun 1883, 7.
51. *Straits Times*, 20 Mar 1884, 2.
52. 'The Straits Settlements and British Malaya' *Straits Times Weekly*, 16 Jul 1884, 9.
53. Liu, 23n1.
54. Ibid.
55. 'No. II', *Straits Times Weekly Issue,* 18 Oct 1884, 9.
56. *Singapore Free Press*, 20 Dec 1884, 211.
57. 'The Year 1884', *Straits Times*, 2 Jan 1885, 2.
58. 'Annual Reports on the Raffles Library and Museum for the Year 1884', *Straits Settlements Annual Department Reports for the Year 1884* (Singapore: Government Printer, 1885), 5.
59. 'Annual Reports on the Raffles Library and Museum for the Year 1885', *Straits Settlements Annual Department Reports for the Year 1885* (Singapore: Government Printer, 1886), 4.
60. *Straits Times Weekly Issue*, 23 Jan 1886, 8.
61. 'Advertisements Column 4', *Straits Times*, 11 Aug 1887, 2.
62. *Straits Times Overland Journal*, 4 Aug 1881, 1.
63. Hanitsch, 551, C1n23.

64. Ibid, 553.
65. Ibid, 554.
66. Robert Ignaz Lendlmayer von Lendenfeld (1858–1913) was an Austrian zoologist. At the time of his application, he had just spent five years in Australia and New Zealand where he taught at the Agricultural College and the Sydney Technical College. During his time there, he conducted extensive studies on the invertebrates of the sea. After failing to secure an appointment in Singapore, von Lendenfeld obtained a position as an assistant at University College, London. He later became Professor at the University of Czernowitz (Chernivtsi in western Ukraine), and then Professor and Director of the Zoological Institute and then Rector at the Karl-Ferdinand University in Prague.
67. Hanitsch, 554, C1n23.
68. 'Annual Report on the Raffles Library and Museum for the Year 1887', *Straits Settlements Annual Department Reports for the Year 1887* (Singapore: Government Printer, 1888), 18.

Chapter 3

1. Between 1857 and 1883, Hume accumulated the largest collection of Asiatic birds in the world. This collection was housed in a museum and library at his home in Rothney Castle in Simla. He made numerous collecting expeditions to collect specimens, including expeditions in North Tenasserim in the 1870s. In 1884, Hume donated his collection of over 75,000 specimens to the British Museum. See *Allan Octavian Hume: Father of the Indian National Congress, 1829–1912* , ed. William Wedderburn (London: T Fisher Unwin, 1913), 39–46.
2. According to Wells, Davison had 'one of the largest personal, or personally supervised specimen tallies in the history of peninsular collecting.' *See* David R. Wells, *The Birds of the Thai-Malay Peninsula: Passerines*, vol. 2 (London: Christopher Helm, 2007).
3. Gretchen Liu, *One Hundred Years of the National Museum, Singapore 1887–1987* (Singapore: National Museum, Singapore, 1987), 24.
4. 'Annual Report on the Raffles Library and Museum for the Year ending 31st December 1888', *Straits Settlements Annual Reports for the Year 1888* (Singapore: Government Printers, 1889), 35.
5. 'Annual Report on the Raffles Library and Museum for the Year ending 31st December 1889', *Straits Settlements Annual Reports for the Year 1889* (Singapore: Government Printers, 1890), 211.
6. 'White Ants in the Museum', *Straits Times Weekly Issue*, 23 Sep 1891, 13.
7. 'Annual Report on the Raffles Library and Museum for the Year Ending 31st December 1891', *Straits Settlements Annual Reports for the Year 1891* (Singapore: Government Printer, 1892), at 37.
8. George Paddison Owen (1856–1928) was a well-known sportsman and hunter. He was a keen tennis player and one of the first members of the Polo Club. For many years, he served as secretary of the Singapore Sporting Club as well as the Singapore Cricket Club. *See* 'Death of Mr G. P. Woen: Noted Sportsman', *Straits Times*, 25 May 1928, 9.

9. 'Annual Report on the Raffles Library and Museum for the Year ending 31st December 1888', *Straits Settlements Annual Reports for the Year 1888* (Singapore: Government Printers, 1889), 36.
10. 'Annual Report on the Raffles Library and Museum for the Year ending 31st December 1889', *Straits Settlements Annual Reports for the Year 1889* (Singapore: Government Printers, 1890), 211.
11. 'The Sudden Death of Mr. Davison', *Straits Times Weekly Issue*, 22 Feb 1893, 2. *Straits Settlements Annual Reports for the Year 1888* (Singapore: Government Printers, 1889).
12. Ibid.
13. Ibid.
14. Kay Gillis & Kevin Y. L. Tan, *The Book of Singapore's Firsts* (Singapore: Singapore Heritage Society & Talisman, 2006), 61.
15. 'Annual Report on the Raffles Library and Museum for the Year ending 31st December 1889', *Straits Settlements Annual Reports for the Year 1889* (Singapore: Government Printers, 1890).
16. Ibid.
17. *Annual Report on the Raffles Library and Museum for the Year ending 31st December 1892* (Singapore: Government Printers, 1893).
18. *Report on the Raffles Library and Museum for the Quarter ending 30th September 1889.*
19. Ibid.
20. Ibid.
21. Ibid.
22. *Annual Report of the Raffles Library and Museum for the Year Ending 31st December 1891* (Singapore: Government Printer, 1892).
23. *Report on the Raffles Library and Museum for the Quarter ending 30th September 1889.*
24. Alfred R. Wallace, 'On the Physical Geography of the Malay Archipelago', *Journal of the Royal Geographical Society* 33 (1863): 227–9.
25. *Report on the Raffles Library and Museum for the Quarter ending 30th September 1889.*
26. 'Local and General', *Straits Times Weekly Issue*, 1 Jul 1891, 2.
27. 'Local and General', *Daily Advertiser*, 31 Mar 1891, 3.
28. *Annual Report of the Raffles Library and Museum for the Year Ending 31st December 1891* (Singapore: Government Printer, 1892).
29. 'Correspondence' *Singapore Free Press*, 25 Jun 1892, 2; see also *Straits Times*, 25 Jun 1892, 2.
30. 'Muar News', *Straits Times Weekly Issue*, 29 June 1892, 9.
31. *Annual Report on the Raffles Library and Museum for the Year ending 31st December 1907* (Singapore: Government Printer, 1908).
32. Ibid.
33. Ibid.
34. 'A Whale of a Gift for Malaysian National Museum', *Straits Times*, 6 May 1974, 8.
35. 'The Sudden Death of Mr Davison', *Straits Times Weekly Issue*, 22 Feb 1893, 2.
36. Ibid.
37. 'The Sudden Death of Mr William Davison', *Straits Times Weekly Issue*, 8 Feb 1893, 5.

38. 'Local and General', *Daily Advertiser*, 21 Jul 1893, 3.
39. Ibid.
40. Liu, 26n3.
41. It was reported that he had assumed his duties on 10 April 1893 'pending confirmation by the Secretary of State.' *See* 'Local & General' *Daily Advertiser*, 10 Apr 1893, 3.
42. *See* Tom Harrison, '"Second to None": Our First Curator (and Others)' (1961) *Sarawak Museum Journal (New Series) 10*, 17–29.
43. 'Annual Report on the Raffles Library and Museum for the Year ending 31st December 1893'. *Straits Settlements Annual Reports for the Year 1893* (Singapore: Government Printer, 1894), 12.
44. Ibid.
45. Ibid.
46. Ibid., 17.
47. Ibid.
48. Ibid., 17–18.
49. Ibid., 18–19.
50. Ibid.
51. 'The Raffles Library', *Daily Advertiser*, 24 Jan 1894, 2.
52. 'Annual Report on the Raffles Library and Museum for the Year ending 31st December 1894', *Straits Settlements Annual Reports for the Year 1894* (Singapore: Government Printer, 1895), 15.
53. Liu, 27n2.
54. 'Mainly About Malayans by The Onlooker' *Straits Times*, 18 Aug 1940, 8.
55. Audrey Z Smith, *A History of the Hope Entomological Collections in the University Museum Oxford* (Oxford: Clarendon Press, 1986), 60. Smith had access to the Hanitsch family records through Hanitsch's grandson Richard Hanage.
56. 'Annual Report on the Raffles Library and Museum for the Year ending 31st December 1895', *Straits Settlements Annual Reports for the Year 1895* (Singapore: Government Printer, 1896).
57. Ibid., 13.
58. Ibid.
59. Ibid.
60. Ibid.
61. Ibid.
62. *Annual Report on the Raffles Library and Museum for the Year ending 31st December 1913* (Singapore: Government Printer, 1914).
63. 'Annual Report on the Raffles Library and Museum for the Year ending 31st December 1896', *Straits Settlements Annual Reports for the Year 1896* (Singapore: Government Printer, 1897), 69.
64. Ibid., 72.
65. Ibid.
66. *Annual Report on the Raffles Library and Museum for the Year ending 31st December 1899* (Singapore: Government Printer, 1900).
67. *Annual Report on the Raffles Library and Museum for the Year ending 31st December 1901* (Singapore: Government Printer, 1902).
68. Ibid.
69. *Annual Report on the Raffles Library and Museum for the Year ending 31st December 1902* (Singapore: Government Printer, 1903).

70. Hanitsch, 558, C1n23.
71. 'Annual Report on the Raffles Library and Museum for the Year ending 31st December 1897', *Straits Settlements Annual Reports for the Year 1897* (Singapore: Government Printer, 1898), 18.
72. 'Annual Report on the Raffles Library and Museum for the Year ending 31st December 1899', *Straits Settlements Annual Reports for the Year 1899* (Singapore: Government Printer, 1900), 29.
73. 'Annual Report on the Raffles Library and Museum for the Year ending 31st December 1900', *Straits Settlements Annual Reports for the Year 1900* (Singapore: Government Printer, 1901), 14.
74. 'Annual Report on the Raffles Library and Museum for the Year ending 31st December 1901', *Straits Settlements Annual Reports for the Year 1901* (Singapore: Government Printer, 1902), 5.
75. Ibid.
76. 'Annual Report on the Raffles Library and Museum for the Year ending 31st December 1902', *Straits Settlements Annual Reports for the Year 1902* (Singapore: Government Printer, 1903), 26.
77. Ibid., 29.
78. 'Annual Report on the Raffles Library and Museum for the Year ending 31st December 1904', *Straits Settlements Annual Reports for the Year 1904* (Singapore: Government Printer, 1905), 25.
79. Ibid.
80. 'Annual Report on the Raffles Library and Museum for the Year ending 31st December 1905', *Straits Settlements Annual Reports for the Year 1905* (Singapore: Government Printer, 1906), 4.
81. 'Annual Report on the Raffles Library and Museum for the Year ending 31st December 1907', *Straits Settlements Annual Reports for the Year 1907* (Singapore: Government Printer, 1908), 4.
82. 'Annual Report on the Raffles Library and Museum for the Year ending 31st December 1911', *Straits Settlements Annual Reports for the Year 1908* (Singapore: Government Printer, 1909), 4.
83. 'Annual Report on the Raffles Library and Museum for the Year ending 31st December 1911', *Straits Settlements Annual Reports for the Year 1911* (Singapore: Government Printer, 1912), 7.
84. 'Annual Report on the Raffles Library and Museum for the Year ending 31st December 1911', *Straits Settlements Annual Reports for the Year 1911* (Singapore: Government Printer, 1912), 9.
85. Walter Makepeace (1859–1941) was a journalist, editor and later joint proprietor of the *Singapore Free Press*. He arrived in Malacca in 1884 as a schoolmaster but left the education service in 1887 to become an assistant editor at the *Singapore Free Press*. He rose to become joint proprietor of the newspaper in 1895. In 1918, the Singapore Centenary Committee commissioned Makepeace to put together an official history of the colony. Makepeace was lead editor of the massive two-volume work and was assisted by Gilbert E. Brooke, Chief Health Officer, and Sir Roland St John Braddell, a well-known lawyer, Municipal Commissioner and member of the Raffles Library and Museum Committee. The book was published as *One Hundred Years of Singapore*, 2 vols, ed. Walter Makepeace, Gilbert E. Brooke & Roland St John Braddell (London: John Murray, 1921).

86. 'Annual Report on the Raffles Library and Museum for the Year ending 31st December 1909', *Straits Settlements Annual Reports for the Year 1909* (Singapore: Government Printer, 1910), 5.

87. 'Annual Report on the Raffles Library and Museum for the Year ending 31st December 1912', *Straits Settlements Annual Reports for the Year 1912* (Singapore: Government Printer, 1913), 3.

88. Hanitsch, 566, C1n23.

89. 'Annual Report on the Raffles Library and Museum for the Year ending 31st December 1907', *Straits Settlements Annual Reports for the Year 1907* (Singapore: Government Printer, 1908), 4.

90. 'Annual Report on the Raffles Library and Museum for the Year ending 31st December 1908', *Straits Settlements Annual Reports for the Year 1908* (Singapore: Government Printer, 1909), 5.

91. 'Annual Report on the Raffles Library and Museum for the Year ending 31st December 1913', *Straits Settlements Annual Reports for the Year 1913* (Singapore: Government Printer, 1914), 3.

92. Hanitsch, 556, C1n23.

93. 'Annual Report on the Raffles Library and Museum for the Year ending 31st December 1909', *Straits Settlements Annual Reports for the Year 1909* (Singapore: Government Printer, 1910), 4.

94. 'Annual Report on the Raffles Library and Museum for the Year ending 31st December 1912', *Straits Settlements Annual Reports for the Year 1912* (Singapore: Government Printer, 1913), 3.

95. 'Annual Report on the Raffles Library and Museum for the Year ending 31st December 1913', *Straits Settlements Annual Reports for the Year 1913* (Singapore: Government Printer, 1914), 1.

96. Ibid.

97. 'Annual Report on the Raffles Library and Museum for the Year ending 31st December 1914', *Straits Settlements Annual Reports for the Year 1914* (Singapore: Government Printer, 1915), 1.

98. *Annual Report on the Raffles Library and Museum for the Year ending 31st December 1918* (Singapore: Government Printer, 1919).

99. 'Annual Report on the Raffles Library and Museum for the Year ending 31st December 1914', *Straits Settlements Annual Reports for the Year 1914* (Singapore: Government Printer, 1915), 2.

100. Liu, 38n3.

101. 'Annual Report on the Raffles Library and Museum for the Year ending 31st December 1918', *Straits Settlements Annual Reports for the Year 1918* (Singapore: Government Printer, 1919), at 3.

102. Ibid.

103. Ibid.

104. Walter Makepeace, Gilbert E Brookes & Roland St John Braddell, *One Hundred Years of Singapore*, 2 vols (London: John Murray, 1921).

105. *See* Panikos Panayi, 'German Immigrants in Britain, 1815–1914' in *Germans in Britain Since 1500*, ed. Panikos Panayi (London: Hambledon Press, 1996) 73–94.

106. Liu, 39n3.

107. Ibid.

108. Bashford Dean, 'Notes on Asiatic Museums', *The Popular Science Monthly* (1907): 484–5.
109. Ibid., at 485.
110. 'Entomology Expert Dies: Rare Distinction from Oxford', *Straits Times*, 14 Aug 1940, 10.
111. Henry N. Ridley, 'An Expedition to Christmas Island', *Journal of the Straits Branch of the Royal Asiatic Society* 45 (1906): 137–55.

Chapter 4

1. 'Social and Personal' *Straits Times*, 12 Jul 1919, 8.
2. *Annual Report on the Raffles Museum and Library for the Year ending 31st December 1919* (Singapore: Government Printer, 1920).
3. Ibid.
4. *Annual Report on the Raffles Museum and Library for the Year ending 31st December 1920* (Singapore: Government Printer, 1922).
5. Ibid. This piece of property later became the site of the first National Library Building in Stamford Road.
6. *Annual Report*, ibid.
7. Ibid.
8. Ibid.
9. Ibid.
10. Ibid.
11. Ibid.
12. Ibid.
13. Ibid.
14. Ibid.
15. Ibid.
16. Ibid.
17. Ibid. This later became the *Journal of the Malayan Branch, Royal Asiatic Society* (1923).
18. *Annual Report on the Raffles Museum and Library for the Year ending 31st December 1920* (Singapore: Government Printer, 1922).
19. Ibid.
20. *Annual Report on the Raffles Museum and Library for the Year ending 31st December 1921* (Singapore: Government Printer, 1922).
21. Ibid.
22. Sir John Alexander Strachey Bucknill, *The Birds of Surrey* (London, 1900).
23. Sir John Alexander Strachey Bucknill & Frederick Nutter Chasen, *The Birds of Singapore Island* (Singapore: WT Cherry, Government Printer, 1927).
24. *Annual Report on the Raffles Museum and Library for the Year ending 31st December 1921* (Singapore: Government Printer, 1922).
25. Ibid.
26. *Annual Report on the Raffles Museum and Library for the Year ending 31st December 1922* (Singapore: Government Printer, 1923).
27. Ibid.
28. Ibid.
29. *Annual Report on the Raffles Museum and Library for the Year ending 31st December 1926* (Singapore: Government Printer, 1927).

30. 'Wray, Leonard, 1852–1942', Royal Commonwealth Society Photographers Index.
31. 'Rimba', *Bygone Selandor: A Souvenir* (Kuala Lumpur, 1922), 74–5.
32. For a brief history of the museum, *see* J. M. Gullick, *A History of Kuala Lumpur 1857–1939*, MBRAS Monograph No 29 (Kuala Lumpur: MBRAS, 2000), 168–9n39. *See also* J. M. Gullick, *History from the Selangor Journal 1892–1897*, MBRAS Reprint No 26 (Kuala Lumpur: MBRAS, 2007), 443–53.
33. *Singapore Free Press*, 23 Sep 1926, 8.
34. For this section, I have relied heavily on information kindly provided by Dr David R Wells in an email to me dated 10 Sep 2014.
35. Ibid.
36. Interview with David Wells, 20 Jan 2014.
37. *See* E. Banks, 'Obituary: Cecil Boden Kloss', *Bulletin of the Raffles Museum* 23 (1950): 336–46.
38. C Boden Kloss, *In the Andamans and Nicobars: The Narrative of a Cruise in the Schooner 'Terrapin'* (London: John Murray, 1903).
39. *Eastern Daily Mail*, 5 Sep 1905, 2.
40. J. E. Hill, 'The Robinson Collection of Malaysian Mammals' *Bulletin of the Raffles Museum* 29 (1960), 1–112.
41. *Annual Report on the Raffles Museum and Library for the Year Ending 31st December 1927* (Singapore: Government Printer, 1928).
42. Ibid, at 1.
43. *Annual Report on the Raffles Museum and Library for the Year ending 31st December 1923* (Singapore: Government Printer, 1924).
44. Ibid.
45. *See* generally, E Banks, 'Obituary: Cecil Boden Kloss' (1950) 23 *Bulletin of the Raffles Museum* 336–346, at 338.
46. Ibid., at 337.
47. Pendlebury to Riley, 7 Jan 1939, DF 306-0009, Natural History Museum Archives, London.
48. *See* M. W. F. Tweedie, 'Obituary: Frederick Nutter Chasen', *Bulletin of the Raffles Museum* 18 (1946): 170–6.
49. F. N. Chasen, *A Handlist of Malaysian Birds: A Systematic List of the Birds of the Malay Peninsula, Sumatra, Borneo and Java, including adjacent small islands* (Singapore: WT Cherry, Government Printer, 1935).
50. F. N. Chasen, *A Handlist of Malaysian Mammals: A Systematic List of Mammals of the Malay Peninsula, Borneo and Java, Including Adjacent Small Islands* (Singapore: WT Cherry, Government Printer, 1940).
51. For a list of his publications, *see* Tweedie, 'Obituary: Frederick Nutter Chase', 170–6.
52. *Annual Report on the Raffles Museum and Library for the Year ending 31st December 1934* (Singapore: Government Printer, 1935).
53. *Annual Report on the Raffles Museum and Library for the Year ending 31st December 1935* (Singapore: Government Printer, 1936).
54. Ibid.
55. *Annual Report on the Raffles Museum and Library for the Year ending 31st December 1937* (Singapore: Government Printer, 1938).
56. *Annual Report on the Raffles Museum and Library for the Year ending 31st December 1939* (Singapore: Government Printer, 1940).

57. *Bulletin of the Raffles Museum* 11 (1935).
58. *Bulletin of the Raffles Museum* 15 (1940).
59. Kitti Thonglongya, 'The History of Mammalogy in Thailand', *Natural History Bulletin of the Siam Society* 25 (1975): 58.

Chapter 5

1. E. J. H. Corner, *The Marquis: A Tale of Syonan-to* (Singapore: Heinemann Asia, 1981) [Hereinafter '*The Marquis*'].
2. John K. Corner, *My Father in His Suitcase: In Search of EJH Corner the Relentless Botanist* (Singapore: Landmark Books, 2013), 122–182 [hereinafter 'J. K. Corner'].
3. 'Events Affecting Raffles Museum & Library from Feb 1942–1945', MSA 1141/249.
4. Ibid.
5. E. J. H. Corner, *Report of Raffles Museum and Library, Singapore*, 15 Sep 1945, 1.
6. Ibid, 5.
7. *The Marquis*, 23n1.
8. Ibid.
9. Ibid.
10. Ibid., 24.
11. Ibid.
12. C. M. Turnbull, *A History of Modern Singapore 1819–2005* (Singapore: NUS Press, 2009) at 202.
13. This building, built between 1926 and 1929, served as the Municipal Building from 1929 till 1951 when it was renamed City Hall. In 1988, it became part of the Supreme Court of Singapore. In 2015, it will open as the new National Gallery of Singapore.
14. *The Marquis*, n 1 above, at 26.
15. *The Marquis*, n 1 above at 28.
16. Hideo Takanadate (1884–1951) was a Japanese geologist and geographer. Born in Fukuoka-Machi, Iwate Prefecture in 1884, he was adopted by Professor Aikitsu Tanakadate, the doyen of Japanese physicists in 1916. At the time of his death, he was Professor at Hosei University. See E. J. H. Corner, 'Prof H Tanakdate', *Nature* 167 (1951): 586–7.
17. Hideo Tanakadate, 'Conservation of a Culture: A Month's Travel in Singapore—Part I', *Asahi Shinbun*, 4 Apr 1942 (in Japanese), translated and published in David J Mabberley, 'A Tropical Botanist Finally Vindicated', *Gardens Bulletin Singapore* 52 (2000): 1–4, 3.
18. *The Marquis*, 31n1.
19. Ibid., 32.
20. Ibid.
21. Ibid., 36.
22. Ibid., 35.
23. E. J. H. Corner, *Report*, 2–3.
24. 'The Singapore Botanic Gardens During 1941–46', *The Gardens' Bulletin Singapore* 11, no. 4 (1947): 264.
25. *The Marquis*, 33n1.
26. Ibid., 7–18.
27. Ibid., 119.

28. The first edition of this book states that it was published by the Syonan-Hakubutsukan (Syonan Museum) and has a publication date of 1941.

29. Ibid., 120.

30. E. J. H. Corner, 'Japanese Men of Science in Malaya during Japanese Occupation', *Nature* 158. no. 4002 (13 Jul 1946); *see also* 'Japanese Published British Book', *Straits Times*, 28 Aug 1946, 3.

31. The Monuments, Fine Arts and Archives programme was established under the Civil Affairs and Military Government sections of the Allied enemies in 1943 to protect cultural property of reoccupying forces. Members of MFAA were known as Monuments Men. *See* Robert M. Edsel, *Monuments Men: Allied Heroes, Nazi Thieves and the Greatest Treasure Hunt in History* (New York: Center Street, 2009).

32. E. J. H. Corner, *Report of Raffles Museum and Library, Singapore*, 15 Sep 1945, 6.

33. Ibid.

34. Archey to HQ BMA (MA), MSA 1141/256, Sep 1945.

35. Ibid.

36. Holttum had returned to England on 18 September 1945 and returned to duty as Director of the Gardens in May 1946. *See* 'The Singapore Botanic Gardens During 1941–46', *The Gardens' Bulletin Singapore* 11 no. 4 (1947): 265.

37. Archey to HQ BMA(M), MSA 1141/256, 26 Sep 1945.

38. Hone to Archey, MSA 1141/256, 31 Oct 1945.

39. Archey to Hone, MSA 1141/256, 6 Nov 1945.

40. Archey to 'C' Division, HQ BMA(M), Kuala Lumpur, MSA 1141/256, 7 Mar 1946.

41. 'Passengers on Empress of Australia' *Straits Times*, 22 Jun 1946, 3.

42. 'Statement of MWF Tweedie, formerly 115552 Flying Officer, RAF, at present Director, Raffles Museum, Singapore', MSA 1143/319, 19 Aug 1946.

43. C. A. Gibson-Hill, 'Bird and Mammal Type Specimens Formerly in the Raffles Museum Collections', *Bulletin of the Raffles Museum* 19 (1949):133–4.

44. Ibid., 134.

45. Chasen to Foenander, MSA 1141/187, 10 Jan 1940.

46. Chasen to Foenander, MSA 1141/187, 15 Jan 1940.

47. 'Six Paris of Elephants' Tusks Loaned to Raffles Museum', *Straits Times*, 17 Mar 1940, 11.

48. Foenander to Director of the Museum in Syonan-to, MSA 1141/187, 11 Sep 1942.

49. Tanakadate to Foenander, MSA 1141/187, 17 Sep 1942.

50. Tanakadate to Foenander, MSA 1141/187, 22 Sep 1942.

51. Archey to Foenander, MSA 1141/187, 4 Dec 1945.

52. Foenander to Director, Raffles Museum, MSA 1141/220, 21 Oct 1947.

53. Gibson-Hill to Foenander, MSA 1141/220, 23 Oct 1947.

54. Foenander to Gibson-Hill, MSA 1121/220, 28 Oct 1947.

Chapter 6

1. When the Japanese attacked Malaya in 1941, Tweedie joined the Royal Air Force as a Pilot Officer.

2. M. W. F. Tweedie, 'The Relation of a Proposed Malayan University to the Raffles Museum and Library, Singapore', MSA 1142/298, 2 Sep 1948.

3. Ibid.

4. H. C. Lepper, 'Memorandum', MSA 1142/281, 6 Dec 1946.

5. Gibson-Hill to Secretary for Economic Affairs, MSA 1142/281, 6 Nov 1947.
6. Annual Report on the Raffles Library and Museum for the Year ending 31st December 1949 (Singapore: Government Printer, 1950).
7. *Annual Report on the Raffles Library and Museum for the Year ending 31st December 1948* (Singapore: Government Printer, 1949).
8. Ibid.
9. Ibid.
10. *See* Diane McNichols, Edwin Clarence Tweedie & Ellen Wyckoff, *The Tweedie Family: A Genealogy* (2005), 83.
11. Peter K. L. Ng and C. M. Yang, 'On Michael Wilmer Forbes Tweedie', *Raffles Bulletin of Zoology* 37 no. 1 & 2 (1989): 160.
12. M. W. F. Tweedie, 'Zoology in the Raffles Museum, 1932–1957' in *Our Heritage: Zoological Reference Collection: Official Opening Souvenir Brochure*, 31 Oct 1988, 2.
13. Ibid., 3.
14. Ng and Yang, 161n11.
15. Ibid.
16. Tweedie, 'Zoology in the Raffles Museum, 1932–1957', 2–3.
17. This brief biographical account of Gibson-Hill is based on 'Carl Alexander Gibson-Hill: A Brief Biography' in *An Orientalist's Treasure Trove of Malaya and Beyond: Catalogue of the Gibson-Hill Collection at the National Library Singapore* ed. Irene Lim & comp. Bonny Tan, (Singapore: National Library Board Singapore, 2008) 12–27; and Bonny Tan, 'Carl Alexander Gibson-Hill', *Singapore Infopedia*.
18. John Lisle, *Warwickshire* (London: F Muller, 1936).
19. This volume was later commercially published by the Malayan Nature Society in 1956. Madoc later became Head of Special Branch, Malaya.
20. C. A. Gibson-Hill, 'Notes on the sea Birds Breeding in Malayan Waters', *Bulletin of the Raffles Museum* 23 (1950): 5–64.
21. Interview with Eric Alfred, 20 Jan 2014.
22. Annual Report on the Raffles Library and Museum for the Year ending 31st December 1950 (Singapore: Government Printer, 1951).
23. Interview with Eric Alfred, 20 Jan 2014.
24. Ibid.
25. Ibid.
26. Annual Report on the Raffles Library and Museum for the Year ending 31st December 1954 (Singapore: Government Printer, 1955).
27. Annual Report on the Raffles Library and Museum for the Year ending 31st December 1955 (Singapore: Government Printer, 1956).
28. Ibid.
29. Annual Report on the Raffles Library and Museum for the Year ending 31st December 1956 (Singapore: Government Printer, 1957).
30. Annual Report on the Raffles Library and Museum for the Year ending 31st December 1950 (Singapore: Government Printer, 1951).
31. Annual Report on the Raffles Library and Museum for the Year ending 31st December 1952 (Singapore: Government Printer, 1953).
32. These were: Raffles National Library Bill; Raffles Museum Bill and Botanic Gardens Bill, passed as the Raffles Museum Ordinance (No. 30 of 1957); the Raffles National Library Ordinance (No. 31 of 1957); and the Botanic Gardens Ordinance (No. 32 of 1957).

33. Ordinance No. 30 of 1957.
34. Ordinance No. 31 of 1957, Singapore Statutes. In December 1960, the word 'Raffles' was dropped and the library simply became known as the National Library.
35. *See* the Raffles National Museum (Change of Name) Ordinance, No. 67 of 1960; and 'Out goes the name Raffles', *Straits Times*, 21 Nov 1960, at 9; and 'Raffles' name will live on: Rajaratnam', *Straits Times*, 1 Dec 1960 7.
36. Singapore Legislative Assembly Debates, Vol. 13, 20 July 1960, col. 24.
37. See 'Museum as aid to education', *Straits Times*, 6 Feb 1961, 4.
38. Ibid.
39. 'Singapore's museum in Bad Shape, says a Raffles Descendant', *Straits Times*, 1 Jul 1965, 7.
40. 'Singapore's Museum Second to None in Malaysia', *Straits Times*, 30 June 1965, 9.
41. Ibid.
42. 'National Museum Proposals for Inclusion in the 2nd Developmental Plan (1966–1970), Ministry of Culture files, MC165-71 Pt 6, 18 May 1965.
43. Eric R Alfred, 'Proposal under Appendix C of Economic Development Circular No 161 dated 10 Apr 1961, Ministry of Culture, MC165-71 Pt 6, 11 Jun 1965.
44. '$470,000 plan to renovate the National Museum', *Straits Times*, 31 December 1967, 16.
45. President's Address, *Singapore Parliamentary Reports*, Vol. 27, 6 May 1968, col. 21.
46. 'The New Cabinet', *Straits Times*, 17 April 1968, 10.
47. 'National Museum', *Singapore Parliamentary Reports*, Vol. 28, 13 Dec 1968, col. 208.
48. Ibid.
49. *Singapore Parliamentary Reports*, Vol. 28, 13 Dec 1968, col. 208.
50. 'UNESCO Experts to Help Set Up Science Museum', *Straits Times*, 1 Nov 1968, 13.
51. *See* 'A Change for the National Museum', *Straits Times*, 14 Dec 1968, 7.
52. Eric R. Alfred, 'Plan for the Reorganisation of the National Museum, Singapore as a Natural History Museum', Ministry of Culture, MC165-71 Pt 6, 29 Aug 1969.
53. 'Plan to Give the Museum a New "Live" Look', *Straits Times*, 16 Sep 1970, 5.
54. '$2 Million Plan to Turn the Museum into Exciting "Living" Display', *Straits Times*, 15 Feb 1970, 7.
55. 'Plan to Give the Museum a New 'Live' Look', *Straits Times*, 16 Sep 1970, 5.
56. '$5m Science Centre to Go Up Soon', *Straits Times*, 12 Mar 1970, 5.
57. 'S'pore Plans to Get Support From UNESCO', *Straits Times*, 25 Apr 1970, 9.
58. 'Plan for Two Divisions at $5.2 Mil Science Centre', *Straits Times*, 8 Jul 1970, 8.
59. Ibid.
60. *See* the Science Centre Act (Cap 286), Singapore Statutes. This was passed on 25 Sep 1970.
61. 'Science Centre May Attract 340,000 Visitors Each Year', *Straits Times*, 23 July 1970, 11.
62. *Singapore Parliamentary Debates*, Vol. 30, 22 July 1970, cols. 137–8.
63. Ibid., at col. 142.
64. 'Plan for Two Divisions at $5.2 Mil Science Centre', *Straits Times*, 8 July 1970, 8.
65. 'Science Centre at Jurong', *Straits Times*, 3 December 1970, at 22.
66. 'Wee to head Science Centre Board for three years', *Straits Times*, 30 Nov 1970, 21.
67. Chia Poteik 'Security and defence again get the lion's share', *Straits Times*, 9 Mar 1971, 6.

68. 'Science Centre at Jurong', *Straits Times*, 3 Dec 1970, 22.
69. 'Science Centre Design Contest', *Straits Times*, 23 Feb 1971, 9.
70. 'Architectural Design Prize Winners', *Straits Times*, 28 Nov 1971, 7.
71. 'Culture Ministry will Take Over Museum', *Straits Times*, 31 Mar 1972, 12.
72. 'Displays Not Meant To Trace History of S'pore', *Straits Times*, 27 Dec 1972, 9.
73. Ibid.
74. 'Schools Science Centre Now at Alexandra', *Straits Times*, 13 Oct 1972, 11.
75. 'Culture Ministry Will Take Over Museum', *Straits Times*, 31 Mar 1972, 12.
76. 'Only The Best People to Run Museum', *Straits Times*, 18 Nov 1972, 14.
77. Interview with Eric Alfred, 20 Jan 2014.
78. 'When a Beauty Meets The Beasts', *Straits Times*, 15 Jul 1976, 32.

Chapter 7

1. The other members were Sng Yew Chong, Rex Shelley and Ronald Sng.
2. Interview with Bernard Tan, 3 Jun 2014.
3. Ibid.
4. Dixon to Pope, 9 Jul 1971, DF206/137, Museum of Natural History Archives, London.
5. 'Type' or 'Holotype' specimens refer to the physical examples of an organism first used to describe the new species. There are also 'Syntypes' (one or more biological types listed in the description where no holotype has been designated) and 'Paratypes' (specimens of a type series other than the holotype). For example, the Zoological Reference Collection had, as at 2001, 48 'types' in its herpetological (amphibian) collection including 9 holotypes, 27 paratypes, and 12 syntypes. *See* Indraneil Das & Kelvin K. P. Lim, 'Catalogue of Herpetological Types in the Collection of the Raffles Museum of Biodiversity Research, National University of Singapore', *The Raffles Bulletin of Zoology* 49 no. 1 (2001): 7–11.
6. This 'Lord Cranbrook' refers to John David Gathorne-Hardy (1900–78), Fourth Earl of Cranbrook who was Trustee of the British Museum between 1964 and 1973. His son, Dr Gathorne Gathorne-Hardy (b. 1933) was the 'Lord Medway' referred to in his correspondence. Medway succeeded his father as the 5th Earl of Cranbrook in 1978. Medway served as Assistant at the Sarawak Museum between 1956 and 1958, and was later Senior Lecturer in Zoology at the University of Malaya in Kuala Lumpur from 1961 to 1971. *See* G. W. H. Davison, Hoi Sen Yong & D. R. Wells, 'Cranbrook at Eighty: His Contributions So Far—Ornithologist, Mammalogist, Zooarchaeologist, Chartered Biolgist and Naturalist', *The Raffles Bulletin of Zoology* 1–7 (2013), supplement no. 29.
7. A. P. Coleman, 'Memorandum', 12 Aug 1971, DF206/137, Museum of Natural History Archives, London.
8. J. P. Harding (Keeper of Zoology) to Sir Frank Claringbull (Director, Museum of Natural History), 17 Aug 1971, DF206/137, Museum of Natural History Archives, London.
9. Berry to Lord Cranbrook, 22 Oct 1971, DF206/137, Museum of Natural History Archives, London.
10. 'Dr Tham Ah Kow' obituary, *Straits Times* 22 Oct 1987, 30.
11. 'Promotions for 21 Senior Staff at Varsity', *Straits Times*, 6 Feb 1971, 3.
12. 'S'pore Asks UNESCO Aid', *Straits Times*, 4 Feb 1967, 5.

13. Raoul Sèrene (1909–80) was a famous French carcinologist who was UNESCO consultant. He was based in Singapore for many years. *See* Martyn E. T. Low, S. H. Tan and Peter K. L. Ng, 'The Raffles Bulletin, 1928–2009: Eight Decades of Brachyuran Crab Research (Crustacea: Decopoda), *The Raffles Bulletin of Zoology*, supplement no. 20.(2009): 291–307

14. Raoul Sèrene, 'What a Reference Collection Is', Ministry of Culture, MC 167/71 Pt 3, 18 Feb 1969.

15. Sèrene to Tham, Ministry of Culture, 2 Jan 1970, MC165-71 Pt 6, Ministry of Science & Technology, National Archives, Singapore.

16. Ibid.

17. Tham to Hoe, Ministry of Culture, 11 Apr 1970, MC165-71 Pt 6, Ministry of Science & Technology, National Archives, Singapore.

18. Tham Ah Kow (Director, Fisheries Biology Unit, Department of Biology, University of Singapore) to Au Yee Pan (Principal Assistant Secretary, Ministry of Science and Technology) MS 639/68/II, 2 May 1970

19. Hoe to Tham, 5 Mar 1971, MC165-71 Pt 6, Ministry of Science & Technology, National Archives, Singapore.

20. Tham to Quahe, 9 Mar 1971, MC165-71 Pt 6, Ministry of Science & Technology, National Archives, Singapore.

21. Hoe to Quahe, 30 Mar 1971, MC165-71 Pt 6, Ministry of Science & Technology, National Archives, Singapore.

22. Lim to Quahe, 17 Sep 1971, MC165-71 Pt 6, Ministry of Science & Technology, National Archives, Singapore.

23. Nan Elliot was Professor of Physiology at Nanyang University. In 1980, Nanyang University was to merge with the University of Singapore to become the National University of Singapore.

24. Berry to Gathorne, the Lord Medway, 22 Apr 1972, DF206/137, Museum of Natural History Archives, London.

25. Gathorne, the Lord Medway to Chuang, 28 Apr 1978, DF206/137, Museum of Natural History Archives, London.

26. Marshall to Jackman, 14 Mar 1973, MC165-71 Pt 6, Ministry of Science and Technology, National Archives, Singapore.

27. Ibid.

28. Amadon to J. F. Conceicao (Chairman, National Museum Board), 18 May 1973, MC165-71 Pt 6, Ministry of Science and Technology, National Archives, Singapore.

29. Timothy P. Barnard, 'The Raffles Museum and the Fate of Natural History in Singapore' in *Nature Contained: Environmental Histories of Singapore*, ed. Timothy P Barnard, (Singapore: NUS Press, 2014), 202 [hereinafter 'Barnard'].

30. '$5m Property Takeover', *Straits Times* 1 Mar 1973, 28.

31. Johnson to Koh, 12 Apr 1972, MC165-71 Pt 6, Ministry of Science & Technology, National Archives, Singapore.

32. Tham Ah Kow, File Note, 13 Apr 1972, MC165-71 Pt 6, Ministry of Science & Technology, National Archives, Singapore.

33. When the RMBC closed down in April 1978, the samples were shipped to the Marine Biological Center, Tokai University, Japan. Interview with C. M. Yang, 16 Jan 2014.

34. Interview with C. M. Yang, 16 Jan 2014.

35. Ibid.

36. Interview with Eric Alfred, 20 Jan 2014.
37. The collection was made up of 15,000 mammals, 31,000 birds, 5,000 reptiles and frogs, 12,000 fishes, 10,000 crabs, 18,000 molluscs, 10,000 insects, 25,000 other invertebrates. Interview with C. M. Yang, 16 Jan 2014.
38. Maritime and Port Authority of Singapore.
39. Interview with C. M. Yang, 16 Jan 2014.
40. In 1987, these books were transferred to the Science Library.
41. Ibid.
42. Ibid.
43. Ibid.
44. Barnard, 207n33.

Chapter 8

1. F. J. Ebling, 'Memorandum', July 1977, MC165-71 Pt 6, Ministry of Science & Technology, National Archives, Singapore.
2. Wells had obtained his PhD in 1966 at the School of Biological Sciences at the University of Malaya for his thesis, *Breeding Seasons: Gonad Cycles and Mount in Three Malayan Munias.*
3. Interview with David R Wells, 20 Jan 2014.
4. Ibid.
5. Ibid.
6. Wells to Short, 10 Sep 1977, on file with RMBR.
7. David W Snow (Sub-Dept of Ornithology, British Museum) to David Wells, 27 Sep 1977, DCF206/137, Museum of Natural History Archives, London.
8. Galbraith to Yang, 12 Dec 1977, DCF206/137, Museum of Natural History Archives, London.
9. Short to Yang, 21 Sep 1977, DCF206/137, Museum of Natural History Archives, London.
10. 'Estimated Values of the Specimens of the Zoological Reference Collection' on file with RMBR (undated).
11. Barnard, 206n29.
12. Interview with C. M. Yang, 3 Nov 2014.
13. Barnard, 207.
14. Roland E Sharma, 'The Zoological Reference Collection: The Interim Years 1972–1980' in *Our Heritage: Zoological Reference Collection*, Official Opening Souvenir Brochure, 31 Oct 1988 (Singapore: National University of Singapore, Zoological Reference Collection 1988), 11.
15. Interview with Nancy Byramji, 31 May 2014.
16. Sharma, 11.
17. Interview with C. M. Yang,16 Jan 2014.
18. Sharma, 11.
19. Yang to Short, 3 Mar 1979, on file with RMBR.
20. Goelet to Lee Kuan Yew, draft letter, 28 Mar 1979, with RMBR. Paul Sweet of AMNH sent a copy of draft to C. M. Yang.
21. Interview with Peter K. L. Ng, 24 Oct 2014.
22. 'Officials of Nature Society', *Straits Times*, 13 Jul 1977, 7.

23. Interview with P. N. Avadhani, 19 Nov 2014.
24. Interview with C. M. Yang, 3 Nov 2014.
25. Nancy Byramji now goes by the pen name 'Aurora Hammonds'.
26. Her first article on nature was 'Gardens Monkeys on the Rampage' *Straits Times*, 18 Apr 1970, 6.
27. Interview with C. M. Yang, 16 Jan 2014.
28. Nancy Byramji, 'Save Our Heritage: Priceless Raffles Collection May End In The Dustbin Unless $70,000 A Year is Found', *Straits Times*, 29 Apr 1979, 1.
29. Ibid.
30. Ibid.
31. Interview with Nancy Byramji, 31 May 2014.
32. Byramji, 1.
33. Ibid.
34. 'How We Can Save Our Heritage', *Straits Times*, 7 May 1979, 17.
35. Koh Yan Poh, 'Move to Save Century-Old Collection of Specimens' *Straits Times*, 16 May 1979, 8.
36. Interview with C. M. Yang, 16 Jan 2014.
37. Evelyn Ng, 'Nantah harks to the call of rare specimens' *Straits Times*, 16 Aug 1979, 7.
38. 'Talks to keep Raffles Collection intact', *Straits Times*, 13 Jul 1979, 16.
39. 'The heritage savers', *Sunday Times*, 19 Aug 1979, 12.
40. R. E. Sharma, 'The Zoological Reference Collection—A Valuable Heritage', *Campus News*, University of Singapore, 30 Nov 1979.
41. 'History and Accomplishments', Nature Society (Singapore) website <http://www. nss.org.sg/about.aspx?id=2> (accessed 13 July 2014).
42. 'No-Home Threat to Priceless Raffles Collection', *Straits Times*, 27 Nov 1980, 13.
43. Ibid.
44. Letter from Australian High Commission, Singapore, 5 Nov 1980; Canberra Foreign Affairs, 28 Oct 1980.
45. Sharma to Hooi, 15 Dec 1980, MC165-71 Pt 6, Ministry of Science & Technology, National Archives, Singapore.
46. Interview with C. M. Yang, 5 Nov 2014.
47. Sharma to Hooi, 15 Dec 1980, MC165-71 Pt 6, Ministry of Science & Technology, National Archives, Singapore.
48. Interview with C. M. Yang, 3 Nov 2014.

Chapter 9

1. See *Our Heritage: Zoological Reference Collection*, Official Opening Souvenir Brochure, 31 Oct 1988 (Singapore: National University of Singapore, Zoological Reference Collection 1988).
2. Mees to C. M. Yang, 19 Oct 1988, on file with RMBR.
3. Cranbrook to C. M. Yang, 6 Aug 1988, on file with RMBR.
4. V. Weitzel, C. M. Yang & C. P. Groves, 'A Catalogue of Primates in the Singapore Zoological Reference Collection; Department of Zoology; National University of Singapore (Formerly the Zoological Collection of Raffles Museum)', *Raffles Bulletin of Zoology* 36, no. 1: (1988) 1–166.

5. D. A. Polhemus & J. T. Polhemus, 'The Aphelocheirinae of tropical Asia (Heteroptera: Naucoridae)', *Raffles Bulletin of Zoology* 36, no. 1 (1988): 167–300.

6. Interview with Peter K. L. Ng, 24 Oct 2014.

7. Interview with Lam Toong Jin, 6 Nov 2014.

8. Since 2001, INDECO has been known as Indeco Consulting Group. It is a subsidiary of CPG Corporation (which was the former Public Works Department).

9. At the time of the move, the Raffles Museum storage cases comprised about 100 cabinets for mammal and bird skins, skulls, eggs, insects, molluscs, etc measuring between 3 to 7 feet in height, 4 to 6 feet in width, and 2 to 3 feet in depth; some 300 wooden boxes for small to medium skins measuring between 1 to 2 feet in height, 3 to 4 feet in width, and 1 to 3 feet in depth; and 20 huge wooden crates for large skins, skulls, and skeletons, measuring between 3 to 4 feet in height, 6 to 8 feet in width and 3 feet in depth. Interview with C. M. Yang, 3 Nov 2014.

10. Ibid.

11. According to C. M. Yang, the system required a row of cabinets measuring 36 ft in length with 12 units of cabinets and shelving (each measuring 36 inches W x 18 inches D x 102 inches H).

12. Interview with Lam Toong Jin, 6 Nov 2014.

13. C. M. Yang, 'The Zoological Reference Collection: 1972–1989 Report', *Raffles Bulletin of Zoology* 38, no. 1 (1990): 92–3.

14. H. Morioka & C. M. Yang, 'A Catalogue of Bird Specimens in the Singapore Zoological Reference Collection, Part 1: Struthioniformes—Charadriiformes', *Raffles Bulletin of Zoology,* supplement no. 4 (1996).

15. Interview with Lee Soo Ying, 5 Nov 2014.

16. Interview with Lam Toong Jin, 6 Nov 2014.

17. On file with RMBR.

18. Interview with Peter K. L. Ng, 24 Oct 2014.

19. Lee Soon Yin to Lim Pin, 18 Sep 1998, on file with RMBR.

20. Lynn Seah, 'From Young Scientist to Leading Crab Expert', *Straits Times*, 2 Nov 1995, 2.

21. Interview with Lam Toong Jin, 6 Nov 2014.

22. Theresa Tan, 'Once Upon a Crab', *The AlumNUS* 99 (2014): 23.

23. Peter K. L. Ng, *The Systematics of Sundanian Freshwater Crabs with Comparative Studies on Selected Characters (Crustacea: Brachyura)*, PhD Thesis, Department of Zoology, Faculty of Science, National University of Singapore, 1990.

24. See J. Purseglove & H. M. Burkhill, 'HB Gilliland, 1911-1965: An Appreciation' *Gardens Bulletin, Singapore* (1967): 107–11

25. Peter K. L. Ng to Kevin Tan, 22 Nov 2014, on file with author.

26. In 1990, another curator, Koo Yuan Hsin was hired, but he left after a year.

27. Peter K. L. Ng to Kevin Tan, 22 Nov 2014, on file with author.

28. Ibid.

29. Ibid.

30. Ibid.

31. Ibid. Tan Eng Chye stepped down as Dean in 2007 to become Provost of the University.

32. Peter K. L. Ng, 'Destiny Achieved: A Journey of Discovery', *Straits Times*, 25 Aug 2012, D6.
33. Interview with Peter K. L. Ng, 24 Oct 2014.
34. Peter K. L. Ng, 'Destiny Achieved: A Journey of Discovery', *Straits Times*, 25 Aug 2012, D6.
35. Peter K. L. Ng to Kevin Tan, 22 Nov 2014, on file with author.
36. Ibid.

Chapter 10

1. Jaya Kumar Narayanan, 'Museum Needs More Space, Better Access', *Straits Times*, 2 Jun 2009, 18.
2. Victoria Vaughan, 'Natural History Needs More Room' *Straits Times*, 4 Jun 2009, at B8.
3. Interview with Peter K. L. Ng, 24 Oct 2014.
4. Ibid.
5. Ibid.
6. Tan Dawn Wei, 'Let's Have A Natural History Museum for Singapore', *Sunday Times*, 14 Jun 2009, 26.
7. Peter K. L. Ng, 'Destiny Achieved: A Journey of Discovery', *Straits Times*, 25 Aug 2012, D6.
8. Interview with Leo Tan, 22 May 2014.
9. Interview with Peter K. L. Ng, 24 Oct 2014.
10. Susan Long, 'Many of Science and Dreams' *Straits Times,* 21 Mar 2014, A21.
11. Tan Dawn Wei, '$10m Gift For Natural History Museum. Offer From Unnamed Donor Boosts NUS Bid to Set Up Gallery For Vast Collection', *Sunday Times,* 24 Jan 2010.
12. Interview with Leo Tan, 22 May 2014.
13. Ibid.
14. Interview with Peter K. L. Ng, 24 Oct 2014.
15. Ibid.
16. Interview with Leo Tan, 22 May 2014.
17. Ibid.
18. 'Tan Dawn Wei, 'Museum In Rush to Raise $35m By June', *Straits Times*, 25 Apr 2010.
19. Interview with Leo Tan, 22 May 2014.
20. Ibid.
21. Interview with Peter Ng, 24 Oct 2014.
22. Ibid.
23. Victoria Vaughan, '$46m Raised For Natural History Museum', *Straits Times*, 23 Jul 2010.
24. There was no matching grant for the Singapore Totalisator Board's S$10 million donation since it was considered a government financial institution.
25. Tan Dawn Wei, 'Crab Expert Named Chief Of New Natural History Museum', *Straits Times*, 7 May 2014.
26. Tan Dawn Wei, 'Work Begins on New NUS Museum', *Straits Times*, 12 Jan 2013.

27. Ibid.
28. Interview with Leo Tan, 22 May 2014.
29. Ibid.
30. Ibid.
31. Tan Dawn Wei, 'Museum's $12m Race For Dino Family', *Straits Times,* 10 Jul 2011.
32. Interview with Peter K. L. Ng, 26 Oct 2014.
33. Peter K. L. Ng, 'Destiny Achieved: A Journey of Discovery', *Straits Times,* 25 Aug 2012, D6.
34. Ibid.
35. Ibid.
36. Interview with Leo Tan, 22 May 2014.
37. Interview with Peter K. L. Ng, 24 Oct 2014.

Bibliography

Official Records and Publications
Raffles Library and Museum Annual Reports, 1874–1955
Singapore Legislative Assembly Debates, 1955–65
Singapore Parliamentary Reports, 1965 to date

Correspondences and Minutes
General
Minute by Sir T. S. Raffles on The Establishment of a Malay College at Singapore, 1819
Peter K. L. Ng to Kevin Tan, 22 Nov 2014, on file with author

NATURAL HISTORY MUSEUM ARCHIVES, LONDON
Raffles Museum, Singapore (1928–71) DF 306/38
Raffles Museum, Singapore (1971–84) DF206/137

NATIONAL ARCHIVES OF SINGAPORE
Ministry of Culture, MC165-71
Ministry of Science & Technology, MS 639/68/II

NATIONAL MUSEUM (SINGAPORE)
National Museum files: MSA 1141/187; MSA 1141/249; MSA 1141/256; MSA 1143/319

RAFFLES MUSEUM OF BIODIVERSITY RESEARCH
Cranbrook to C. M. Yang, 6 Aug 1988
'Estimated Values of the Specimens of the Zoological Reference Collection'
Goelet to Lee Kuan Yew, draft letter, 28 Mar 1979
Lee Soon Yin to Lim Pin, 18 Sep 1998
Mees to C. M Yang, 19 Oct 1988
Wells to Short, 10 Sep 1977
Yang to Short, 3 Mar 1979

Laws and Statutes
Botanic Gardens Ordinance (No. 32 of 1957)
Chinese Immigrants Ordinance (No. 3 of 1877)
Raffles Societies Ordinance (No. VII of 1878)
Science Centre Act (Cap 286)
Raffles National Library Ordinance (No. 31 of 1957)
Raffles Museum Ordinance (No. 30 of 1957)
Raffles National Museum (Change of Name) Ordinance (No. 67 of 1960)

Oral History Interview by Author

Eric R Alfred, 20 Jan 2014
Nancy Byramji, 31 May 2014
Lam Toong Jin, 6 Nov 2014
Lee Soo Ying, 5 Nov 2014
Peter Ng Kee Lin, 24 Oct 2014
Bernard Tan Tiong Gie, 3 Jun 2014
Leo Tan Wee Hin, 22 May 2014
David R. Wells, 20 Jan 2014
Yang Chang Man, Mrs, 16 Jan 2014 & 3 Nov 2014

Oral History Interviews (National Archives of Singapore)

Newspapers

Eastern Daily Mail
Singapore Free Press
Straits Observer
Straits Times
Straits Times Overland Journal
Straits Times Weekly Issue

Websites

'Barry, Sir Redmond', *Australian Dictionary of Biography*, National Centre of Biography, Australian National University, published in hardcopy 1969. Available at <http://adb. anu.edu.au/biography/barry-sir-redmond-2946/text4271> (accessed 1 Aug 2014).

'History and Accomplishments', Nature Society (Singapore) website, available at <http:// www.nss.org.sg/about.aspx?id=2> (accessed 13 July 2014).

Tan, Bonny, 'Carl Alexander Gibson-Hill', *Singapore Infopedia*, available at <http:// eresources.nlb.gov.sg/infopedia/articles/SIP_1348_2008-12-01.html> (accessed 15 June 2014).

'Wray, Leonard, 1852–1942', Royal Commonwealth Society Photographers Index, available at <http://www.lib.cam.ac.uk/rcs_photographers/entry.php?id=486> (accessed 1 June 2014).

Books and Book Chapters

Barnard, Timothy P. 'The Raffles Museum and the Fate of Natural History in Singapore'. In *Nature Contained: Environmental Histories of Singapore*, ed. Timothy P. Barnard, 184–211. Singapore: NUS Press, 2014.

Bastin, John. 'Raffles the Naturalist'. In *The Golden Sword: Stamford Raffles and the East*, ed. Nigel Barley, 18–29. London: British Museum Press, 1999.

———. 'William Farquhar: First Resident and Commandant of Singapore'. In *Natural History Drawings: The Complete William Farquhar Collection, Malay Peninsula 1803–1818*, 9–33. Singapore: Editions Didier Millet and National Museum of Singapore, 2010.

Bucknill, Sir John Alexander Strachey. *The Birds of Surrey*. London: R. H. Porter, 1900.

Bucknill, Sir John Alexander Strachey & Frederick Nutter Chasen. *The Birds of Singapore Island*. Singapore: WT Cherry, Government Printer, 1927.

Chasen, Frederick Nutter. *A Handlist of Malaysian Birds: A Systematic List of the Birds of the Malay Peninsula, Sumatra, Borneo and Java, Including Adjacent Small Islands*. Singapore: WT Cherry, Government Printer, 1935.

Cook, J. A. Bethune. *Sir Thomas Stamford Raffles and Some of his Friends and Contemporaries*. London: Arthur H Stockwell, 1918.

Corner, E. J. H. *The Marquis: A Tale of Syonan-to*. Singapore: Heinemann Asia, 1981.

Corner, John K. *My Father in His Suitcase: In Search of EJH Corner the Relentless Botanist*. Singapore: Landmark Books, 2013.

Desmond, Ray, ed. *Dictionary of British and Irish Botanists and Horticulturists*, revised edition. London: Taylor & Francis and the Natural History Museum, 1994.

Gillis, Kay & Kevin Y. L. Tan. *The Book of Singapore's Firsts*. Singapore: Singapore Heritage Society & Talisman, 2006.

Gullick, J. M. *A History of Kuala Lumpur 1857–1939*. MBRAS Monograph No. 29. Kuala Lumpur: MBRAS, 2000.

———. *History from the Selangor Journal 1892–1897*. MBRAS Reprint No. 26. Kuala Lumpur: MBRAS, 2007.

Hanitsch, Richard. 'Raffles Library and Museum'. In *One Hundred Years of Singapore*, vol 1, ed. Walter Makepeace, Gilbert E. Brooke & Roland St John Braddell, 519–66. London: John Murray, 1921.

Kloss, Cecil Boden. *In the Andamans and Nicobars: The Narrative of a Cruise in the Schooner 'Terrapin'*. London: John Murray, 1903.

Lim, Irene, ed. Compiled by Bonny Tan. *An Orientalist's Treasure Trove of Malaya and Beyond: Catalogue of the Gibson-Hill Collection at the National Library Singapore*, 12–27. Singapore: National Library Board Singapore, 2008.

Lisle, John. *Warwickshire*. London: F Muller, 1936.

Liu, Gretchen. *One Hundred Years of the National Museum, Singapore 1887–1987*. Singapore: National Museum, Singapore, 1987.

McNichols, Diane, Edwin Clarence Tweedie & Ellen Wyckoff. *The Tweedie Family: A Genealogy* (2005). Available at < http://www.tweedie.com/TweedieBook.pdf>.

Panayi, Panikos. 'German Immigrants in Britain, 1815–1914'. In *Germans in Britain Since 1500*, ed. Ianikos Panayi, 73–94. London: Hambledon Press, 1996.

Raffles, Sophia. *Memoirs of the Life and Public Services of Sir Thomas Stamford Raffles*. London: John Murray, 1830.

'Rimba'. *Bygone Selangor: A Souvenir*. Kuala Lumpur: C. Grenier, 1922.

Scherren, Henry. *The Zoological Society of London*. London: Cassell & Co, 1905.

Sharma, Roland E. 'The Zoological Reference Collection: The Interim Years 1972–1980'. In *Our Heritage: Zoological Reference Collection, Official Opening Souvenir Brochure*, 31 Oct 1988, 10–11. Singapore: National University of Singapore, Zoological Reference Collection, 1988.

Smith, Audrey Z. *A History of the Hope Entomological Collections in the University Museum Oxford*. Oxford: Clarendon Press, 1986.

Tinsley, Bonny. *Gardens of Perpetual Summer: The Singapore Botanic Gardens*, 31–2. Singapore: National Parks Board & Singapore Botanic Gardens, 2009.

Turnbull, C. M. *A History of Modern Singapore 1819–2005*. Singapore: NUS Press, 2009.

Tweedie, Michael Wilmer Forbes, 'Zoology in the Raffles Museum, 1932–1957'. In *Our Heritage: Zoological Reference Collection: Official Opening Souvenir Brochure, 31*

October 1988, 2–4. Singapore: National University of Singapore, Zoological Reference Collection, 1988.

Wedderburn, William. *Allan Octavian Hume: Father of the Indian National Congress, 1829–1912*. London: T Fisher Unwin, 1913.

Articles

Corner, E. J. H. 'Prof H Tanakdate'. *Nature* 167 (1951): 586–7.

———. 'Japanese Men of Science in Malaya during Japanese Occupation'. *Nature* 63 (1946): 158.

Das, Indraneil & Kelvin K. P. Lim. 'Catalogue of Herpetological Types in the Collection of the Raffles Museum of Biodiversity Research, National University of Singapore'. *The Raffles Bulletin of Zoology* 49, no. 1 (2001): 7–11.

Davison, G. W. H., Hoi Sen Yong & D. R. Wells. 'Cranbrook at Eighty: His Contributions So Far—Ornithologist, Mammalogist, Zooarchaeologist, Chartered Biologist and Naturalist'. *The Raffles Bulletin of Zoology*, supplement no. 29 (2013): 1–7.

Gibson-Hill, C. A. 'Bird and Mammal Type Specimens Formerly in the Raffles Museum Collections'. *Bulletin of the Raffles Museum* 19 (1949): 133–98.

Hill, J. E. 'The Robinson Collection of Malaysian Mammals'. *Bulletin of the Raffles Museum* 29 (1960): 1–112.

Low, Martyn E. T, S. H Tan and Peter K. L. Ng. 'The Raffles Bulletin, 1928–2009: Eight Decades of Brachyuran Crab Research (Crustacea: Decopoda)'. *The Raffles Bulletin of Zoology*, supplement no. 20 (2009): 291–307.

Mabberley, David J. 'A Tropical Botanist Finally Vindicated'. *Gardens Bulletin Singapore* 52 (2000): 1–4.

Morioka, H, & C. M. Yang, 'A Catalogue of Bird Specimens in the Singapore Zoological Reference Collection, Part 1: Struthioniformes–Charadriiformes'. *Raffles Bulletin of Zoology*, supplement no. 4 (1996).

Ng, Peter K. L. & C. M. Yang. 'On Michael Wilmer Forbes Tweedie'. *Raffles Bulletin of Zoology* 37, no. 1 & 2 (1989): 160–7.

Polhemus, D. A & J. T. Polhemus. 'The Aphelocheirinae of tropical Asia (Heteroptera: Naucoridae)'. *Raffles Bulletin of Zoology* 36, no. 1 (1988): 167–300.

Purseglove, J & H. M. Burkhill. 'HB Gilliland, 1911–1965: An Appreciation'. *Gardens Bulletin, Singapore (1967)*: 107–11.

Rookmaaker, Kees. 'Two Former Zoological Gardens in Singapore'. *International Zoo New* 59, no. 5 (2012): 367–72.

Sharma, Rolland E. 'The Zoological Reference Collection—A Valuable Heritage'. *Campus News*, University of Singapore, 30 Nov 1979.

Tanakadate, Hidezo. 'Conservation of a Culture: A Month's Travel in Singapore—Part I'. *Asahi Shinbun*, 4 Apr 1942.

'The Singapore Botanic Gardens During 1941–46'. *The Gardens' Bulletin Singapore* 11, no. 4 (1947): 263–5.

Thonglongya, Kitti. 'The History of Mammalogy in Thailand'. *Natural History Bulletin of the Siam Society* 25 (1975): 53–68.

Tweedie, M. W. F. 'Obituary: Frederick Nutter Chasen'. *Bulletin of the Raffles Museum* 18, (1946): 170–6.

Wallace, Alfred Russel. 'On the Physical Geography of the Malay Archipelago'. *Journal of the Royal Geographical Society* 33 (1863): 217–34.

Weitzel, V, C. M. Yang & C. P. Groves. 'A Catalogue of Primates in the Singapore Zoological Reference Collection; Department of Zoology; National University of Singapore (Formerly the Zoological Collection of Raffles Museum)'. *Raffles Bulletin of Zoology* 36, no. 1 (1988): 1–166.

Yang, C. M. 'The Zoological Reference Collection: 1972–1989 Report'. *Raffles Bulletin of Zoology* 38, no. 1 (1990): 91–116, at 92–3.

Unpublished Dissertations and Working Papers

Ng, Peter K. L. *The Systematics of Sundanian Freshwater Crabs with Comparative Studies on Selected Characters (Crustacea: Brachyura)*. PhD Thesis, Department of Zoology, Faculty of Science, National University of Singapore, 1990.

List of Illustrations

Illustration Credits

Every reasonable effort has been made to trace ownership of copyright materials. The publisher will gladly rectify any errors or omissions.

0.0   Endpaper: Wet Specimens. Images courtesy of Dr Tan Heok Hui, designed by Grace Lin.

0.2   Apollonia. Photo courtesy of Dinosauria International, LLC.

1.6   Singapore Town Hall. Courtesy of National Archives of Singapore.

1.9   Raffles Institution in the 1880s. Courtesy of National Archives of Singapore.

2.3   The only Wallace specimen in the museum. The stuffed bird, an Asian Brown Flycatcher. Photo courtesy of Tan Heok Hui.

2.4   Specimens collected by Wallace from Southeast Asia in his original cabinets. Photo courtesy of Peter K. L. Ng.

2.9   Construction drawings of the proposed Library and Museum at Stamford Road. Collection of the National Museum of Singapore, National Heritage Board.

3.3   The Wallace Line. Reproduced with permission from John van Whye.

3.4   The whale at the Raffles Museum in 1908. Sim Boon Kwang, photo from *Guide to Zoological Collections of the Raffles Museum* (1908).

3.5   Whale at the Labuan Maritime Museum. Photo courtesy of Shih Hse-Te.

3.9   Hanitsch in the curator's office at the museum. Collection of the National Museum of Singapore, National Heritage Board.

3.12   Students from Tampines Primary School. Collection of the National Museum of Singapore, National Heritage Board.

4.6   The old Sarawak Museum building. Tan Kok Kheng Collection, Courtesy of National Archives of Singapore.

4.7   The old Selangor Museum building. Courtesy of National Archives of Singapore.

5.1   Japanese troops marching into Singapore city, 1942. Collection of the Imperial War Museum.

5.2   E. J. H. Corner with his son, John, mid-1941. Image courtesy of John Corner, published in *My Father in His Suitcase* (2013), Landmark Books.

5.3   Japanese staff of the Syonan Museum, 1943. Collection of the National Museum of Singapore, National Heritage Board.

5.4   Corner with the Japanese. Image courtesy of John Corner, published in *My Father in His Suitcase* (2013), Landmark Books.

5.5   Madoc scrutinising a brown booby. Courtesy of Fenella Madoc-Davis.

5.6   Madoc with some locals at the beach. Courtesy of Fenella Madoc-Davis.

5.7   Black-Necked Tailorbird. Courtesy of Fenella Madoc-Davis.

5.8   Black-and-Red Broadbill. Courtesy of Fenella Madoc-Davis.

5.9   Anatomy of a bird. Courtesy of Fenella Madoc-Davis.

6.2   Tweedie visiting the Zoological Reference Collection in 1988. Photo courtesy of Yip Hoi Kee.

6.4   Gibson-Hill and Yusof bin Ishak. Yusof Ishak Collection, Courtesy of National Archives of Singapore.

6.7   Students from Bukit Panjang Government School on excursion to the museum. Bukit Panjang Government School Collection, Courtesy of National Archives of Singapore.

6.8   Christopher Hooi. The Ministry of Information and the Arts Collection, Courtesy of National Archives of Singapore.

6.9   Interior shot of the National Museum, 1964. The Ministry of Information and the Arts Collection, Courtesy of National Archives of Singapore.

6.10   Deer display and elephant skeleton, 1964. The Ministry of Information and the Arts Collection, Courtesy of National Archives of Singapore.

6.11   Sir Solomon Hochoy and Eric Alfred at the National Museum, 1964. The Ministry of Information and the Arts Collection, Courtesy of National Archives of Singapore.

6.12   Vice-President of Cambodian Council Of Ministers and Governor of Cambodian National Bank Son Sann at the National Museum, 1966. The Ministry of Information and the Arts Collection, Courtesy of National Archives of Singapore.

6.13   Existing and proposed plans for the revamped National Museum, 1970. Courtesy of National Archives of Singapore.

6.14   Science Centre in Jurong. Singapore Tourist Promotion Board Collection, Courtesy of National Archives of Singapore.

6.15   Science Centre Logo. Image courtesy of Science Centre Singapore.

7.1   Lecturers and professors of Zoology Department, Bukit Timah Campus, 1969. Image courtesy of Yang Chang Man.

7.2   Common mangrove species, *Episesarma chentongense*. Photo courtesy of Lee Bee Yan.

7.3   Leo Tan and Shirley Lim collecting seashore animals. Photo courtesy of Leo Tan.

7.4   Official opening of the Regional Marine Biological Centre by Toh Chin Chye, 1968. Image courtesy of Yang Chang Man.

7.5   Yang Chang Man. Photo courtesy of Tan Heok Hui.

7.6   Wooden boxes for bird/mammal skins. Image courtesy of Yang Chang Man.

8.2   Lam Toong Jin. Photo courtesy of Yip Hoi Kee.

8.3   Jaws and mounted specimen of Tiger Shark. Image courtesy of Yang Chang Man.

8.4   The museum's famous Black Marlin. Photo courtesy of Yeo Keng Loo.

8.5–8.8 Specimens of the Zoological Reference Collection at Nanyang University. Photos courtesy of Yip Hoi Kee.

8.9   Packing and rearranging specimens while moving to Kent Ridge. Photos courtesy of Yip Hoi Kee.

8.10   University van used for transporting specimens. Photo courtesy of Yip Hoi Kee.

9.1   Science Library Building at Kent Ridge. Photo courtesy of Yip Hoi Kee.

9.2   Opening of the Zoological Reference Collection in 1988. Photo courtesy of Yip Hoi Kee.

9.3   Mounted animals on display at the gallery of the Zoological Reference Collection. Photo courtesy of Yip Hoi Kee.

9.4   Cover of *The Raffles Bulletin of Zoology*. Image courtesy of Yang Chang Mun.

9.5   Unpacking specimens. Photo courtesy of Yip Hoi Kee.

9.6   The many trays at the Dry Collection 178. Photo courtesy of Yip Hoi Kee.

9.7   Broadbills and pittas, some of the 30,000 bird specimens. Photo courtesy of Yip Hoi Kee.

9.8   Wet specimens preserved in 70% ethanol. Photo courtesy of Yip Hoi Kee.

9.9   Mr Tan Teong Hean, with the museum's honorary butterfly curator Mr Khew Sin Khoon, admiring the famous W. A. Fleming Collection of Malayan Butterflies. Photo courtesy of Tan Heok Hui.

9.11   Some of the more than 20,000 plant specimens in the museum's herbarium (SINU). Photo courtesy of Tan Tiang Wah.

9.16   A bird taxidermy class. Photo courtesy of Wang Luan Keng.

9.17   Sir David Attenborough, with staff and students who assisted with the filming of BBC's Life in Cold Blood in Singapore's mangroves in 2008. Photo courtesy of N. Sivosothi.

9.19   Preparing to enter the forests and caves of Gunung Mulu National Park. Photo courtesy of Tan Heok Hui.

9.20   Indonesian fishermen using basket traps at Sebangau in Kalimantan. Photo courtesy of Tan Heok Hui.

9.21   The specimens in the Raffles Museum of Biodiversity Research. Image by Peter K. L. Ng, illustrated by Grace Lin.

10.1   Museum Day 2009 visitors at the original public gallery. Photo courtesy of Tan Heok Hui.

10.2   Some of the local reptiles, preserved in bottles. Photo courtesy of Tan Heok Hui.

10.3   The late Lady McNeice. Photo courtesy of N. Sivosothi.

10.4   Apollonia. Courtesy of Dinosauria International, LLC.

10.8   Artist concept of the new museum facility and the various access points, greenery and bridge to the Alice Lee Plaza in NUS. Image courtesy of W Architects.

10.9   Artist concept of the mangrove garden behind the main building of the new museum. Image courtesy of W Architects.

10.10   The management team of the new museum building at the groundbreaking ceremony on 11 Jan 2013. Photo courtesy of Tan Heok Hui.

10.11   The Ong Tiong Tat & Irene Ong Mammal Zone. Image courtesy of Gsmprjct.

10.12   The Marine Cycle Zone. Image courtesy of Gsmprjct.

11.1   Prince, lit by moonlight. Photo courtesy of Nicklaus Tan.

Index

Related Books by NUS Press

The Annotated Malay Archipelago by Alfred Russel Wallace

John van Wyhe (Editor)

THE MALAY ARCHIPELAGO, the classic account of Victorian naturalist Alfred Russel Wallace's travels through Southeast Asia, first appeared in 1869 and has been much loved by generations of readers ever since. Despite numerous modern reprints with appreciative introductions, this edition is the first—long overdue—fully annotated version to appear in English. The treasure trove of new information it contains illuminates *The Malay Archipelago* like never before.

Through an examination of the historical context, editor John van Whye reveals new aspects of Wallace's life, his sources and the original meanings of this famous book. Following conventions of the time, Wallace often left people, places and publications unidentified, and he referred to most species only by the scientific name current in the 19th century, terms that are unintelligible to most readers today. The explanatory notes, running into the hundreds, provide the common names for species and update their scientific names. People, places and other details that Wallace mentions have been tracked down and identified.

Nature Contained: Environmental Histories of Singapore

Timothy P. Barnard (Editor)

HOW HAS SINGAPORE'S environment—and location in a zone of extraordinary biodiversity—influenced the economic, political, social and intellectual history of the island since the early 19th century? What are the antecedents to Singapore's image of itself as a City in a Garden? Grounding the story of Singapore within an understanding of its environment opens the way to an account of the past that is more than a story of trade, immigration and nation building. Each of the chapters in this volume - focusing on topics ranging from tigers and plantations to trade in exotic animals and the greening of the city, and written by botanists, historians, anthropologists, and naturalists - examines how humans have interacted with and understood the natural environment on a small island in Southeast Asia over the past 200 years, and conversely how this environment has influenced humans. Between the chapters are traveler's accounts and primary documents that provide eyewitness descriptions of the events examined in the text. In this regard, *Nature Contained: Environmental Histories of Singapore* provides new insights into the Singaporean past, and reflects much of the diversity, and dynamism, of environmental history globally.